T0185240

Java Software Development with Event B

Event B

A Practical Guide

Synthesis Lectures on Software Engineering

Editor

Luciano Baresi, *Politecnico di Milano*

The Synthesis Lectures on Software Engineering series publishes short books (75-125 pages) on conceiving, specifying, architecting, designing, implementing, managing, measuring, analyzing, validating, and verifying complex software systems. The goal is to provide both focused monographs on the different phases of the software process and detailed presentations of frontier topics. Premier software engineering conferences, such as ICSE, ESEC/FSE, and ASE will help shape the purview of the series and make it evolve.

Java Software Development with Event B: A Practical Guide
Néstor Cataño Collazos
2020

Model-Driven Software Engineering in Practice: Second Edition
Marco Brambilla, Jordi Cabot, and Manuel Wimmer
2017

Testing iOS Apps with HadoopUnit: Rapid Distributed GUI Testing
Scott Tilley and Krissada Dechokul
2014

Hard Problems in Software Testing: Solutions Using Testing as a Service (TaaS)
Scott Tilley and Brianna Floss
2014

Model-Driven Software Engineering in Practice
Marco Brambilla, Jordi Cabot, and Manuel Wimmer
2012

© Springer Nature Switzerland AG 2022

Reprint of original edition © Morgan & Claypool 2020

All rights reserved. No part of this publication may be reproduced, stored in a retrieval system, or transmitted in any form or by any means—electronic, mechanical, photocopy, recording, or any other except for brief quotations in printed reviews, without the prior permission of the publisher.

Java Software Development with Event B: A Practical Guide
Néstor Cataño Collazos

ISBN: 978-3-031-01422-2 paperback
ISBN: 978-3-031-02550-1 ebook
ISBN: 978-3-031-00340-0 hardcover

DOI 10.1007/978-3-031-02550-1

A Publication in the Springer series
SYNTHESIS LECTURES ON SOFTWARE ENGINEERING
Lecture #5
Series Editor: Luciano Baresi, Politecnico di Milano

Series ISSN   2328-3319 Print   2328-3327 Electronic

Java Software Development with Event B

A Practical Guide

Néstor Cataño Collazos
Google, Inc.

SYNTHESIS LECTURES ON SOFTWARE ENGINEERING #5

ABSTRACT

The cost of fixing software design flaws after the completion of a software product is so high that it is vital to come up with ways to detect software design flaws in the early stages of software development, for instance, during the software requirements, the analysis activity, or during software design, before coding starts. It is not uncommon that software requirements are ambiguous or contradict each other. Ambiguity is exacerbated by the fact that software requirements are typically written in a natural language, which is not tied to any formal semantics. A palliative to the ambiguity of software requirements is to restrict their syntax to boilerplates, textual templates with placeholders. However, as informal requirements do not enjoy any particular semantics, no essential properties about them (or about the system they attempt to describe) can be proven easily. Formal methods are an alternative to address this problem. They offer a range of mathematical techniques and mathematical tools to validate software requirements in the early stages of software development.

This book is a living proof of the use of formal methods to develop software. The particular formalisms that we use are EVENT B and refinement calculus. In short: (i) software requirements as written as User Stories; (ii) they are ported to formal specifications; (iii) they are refined as desired; (iv) they are implemented in the form of a prototype; and finally (v) they are tested for inconsistencies. If some unit-test fails, then informal as well as formal specifications of the software system are revisited and evolved.

This book presents a case study of software development of a chat system with EVENT B and a case study of formal proof of properties of a social network.

KEYWORDS

correct-by-construction, discrete mathematics, EVENT B, formal methods, Java, programming, refinement, software engineering, verification

Contents

Preface

This book is a discourse explaining the software engineering task by means of mathematics. The particular formalism that we use is called predicate calculus. The book starts by introducing predicate calculus objects such as relations and functions and all the underlying operations used to manipulate these objects, giving examples of how they can be composed together to model software. The fundamental reason of using mathematics, and, in particular, predicate calculus to model software is that it allows us to reason about the software we want to build; we are interested in being able to verify if the software we build adheres to certain properties, which are typically condensed in a software requirements document. This is not an easy task and the main motivation behind our actions is to build correct software, software that enjoys those properties described in the software requirements document.

Software testing has had a major impact on software development for the past decades. It has shaped the way people think about the task of software construction. People are not surprised to get to implement software that is infested with bugs. They expect that bugs will be decanted once the software implementation phase finishes. This book uses a different approach; our starting point is that we can and should write software that is correct from the very early stages of software development. Nonetheless, we use testing to animate the software and to check if its running behavior matches its expected behavior.

Our software construction return to mathematics is based on the idea of assigning meaning to software. The idea is to build software the same way that mathematical proofs are conducted, guaranteeing that software product lives up to some intended meaning. The particular mathematical formalism and language we have chosen is called EVENT B, which originated from the Z formalism. Although EVENT B shares essentially the same modeling language for stating state properties as Z, EVENT B and Z offer different modeling mechanisms that are specialized in distinct mathematical aspects. EVENT B and Z are both models for state transition systems. EVENT B's language for expressing the dynamic behavior of state machines is based on events. On the other hand, Z uses a rich schema calculus mechanism for expressing the dynamic behavior of models. Z Schema calculus and EVENT B events coupled with model refinement are different mechanisms. Z also offers refinement, but in practice, Z focuses more on formal specification and EVENT B focuses more on model refinement and coding (whereas it is manually written or tool-generated). Another major difference between EVENT B and Z is in the undertaking and use of invariants. In EVENT B, each event definition produces proof obligations that attest to the correctness of machine invariants.

These invariants might encode safety properties. On the other hand, in Z, invariants are incorporated into the model definitions, altering their meanings. They do not generate proof obligations.

The main motivation for using mathematical formalisms in software development is economical. The cost of fixing software design flaws after the completing of a software product is so high that it is vital to come up with ways to detect software design flaws in the early stages of software development, for instance, during the software requirements, the analysis activity, or during software design, before coding starts. The approach to software construction presented in this book is based on the idea of program refinement and correctness-by-construction.

The book presents a case study of software development of a chat system with EVENT B and a case study of formal proof of properties of a social network. We hope you enjoy the book as much as we enjoyed writing it.

CHAPTER 1

Introduction

This book is a practical guide for the software development of JAVA programs with EVENT B with the EVENTB2JAVA tool. EVENT B is a formal language for the design of systems by the use of discrete mathematics. The EVENT B language is based on set theory and predicate logic. It includes a large set of operations over sets and relations for modeling software systems. It is a simple model for *state machines*, which capture the idea that a system progresses through a set of states by responding to a set of events or actions. Additionally, EVENT B mathematical language permits the definition of invariants, and tool supporting the language can then be used to check if the system attests to the property or not.

This book presents two software examples with EVENT B, a chat application similar to WHATSAPP (https://www.whatsapp.com/) and the *Poporo* social network. For the chat system, we include basic functionality for chatting, forwarding messages, deleting them, and checking if users are actively chatting or not. Poporo is a logical formalisation of a typical social network that includes functionality similar to Facebook's. It additionally includes functionality related to privacy and ownership of social network content.

Chapter 2 introduces EVENT B's syntax, including notations used for sets, relations, and events. A more complex notation is introduced on-the-fly, in the following chapters of the book, as they will require. Chapter 3 presents the analysis and design of the chat system in EVENT B. This is a complete JAVA software development example that goes from software requirements to a close-to-implementation EVENT B model. We use the EVENTB2JAVA JAVA generator (Cataño and Rivera, 2016) to produce JAVA code for the EVENT B model of the chat system. The generated JAVA code is mainly used for validating the correctness of the software requirements of the chat system. The JAVA implementation could be used as a final implementation of the chat system, however, people might attempt to produce a manual implementation that is faster than the code produced by the EVENTB2JAVA JAVA code generator.

Chapter 3 also shows how software requirements can be validated early in the software development process. To validate software requirements, the chapter follows the methodology described in Section 2.4. The methodology (and EVENT B in general) advocates for the construction of software that is *correct-by-construction*. Chapter 3 does not emphasize on the process of making sure that the produced software is correct-by-construction. This endeavor is rather left to Chapter 4. Chapter 4 is dedicated to mathematical proof, so it focuses on the use of particular analysis and techniques to discharge Proof Obligations (POs) that are generated to attest to the correctness of EVENT B models. Proving POs is conducted with the Rodin toolset (Abrial et al., 2010).

This book is targeted to a diverse audience of readers, spanning from Software Engineering professionals with experience in programming but scarce knowledge in Discrete Mathematics to experts in Formal Methods who are looking for real-life formal software development examples. The examples are self-contained. The book is also targeted to academic teachers who want to introduce Discrete Mathematics to undergraduate Computer Science or Software Engineering students, or to students looking for material that supplements their courses. The book assumes the reader has basic programming skills in JAVA or a similar Object-Oriented programming language.

CHAPTER 2

An Overview of EVENT B

In this chapter, we give a broad view of the EVENT B *formal method*. The expression formal method refers to a direct technique for building *dependable* systems. Dependability is the ability of a system to defensively provide a particular service during a period of time. A formal method provides ways to integrate properties into the system design and to mathematically prove system compliance with them. EVENT B is based on Action Systems (Back and Sere, 1991), a formalism describing the behavior of a system by the (atomic) actions that the system carries out. An Action System describes the state space of a system and the possible actions that can be executed in it.

EVENT B software system models are discrete transition system models. EVENT B represents system components as a succession of states connected through a series of transitions (actions) called *events*. States are composed of constants and variables. EVENT B models are composed of *contexts* and *machines*. Contexts define constants, un-interpreted sets, and their properties expressed as `axioms`, while machines define variables and their properties, and state transitions expressed as events. The initialization event computes the initial state of a machine. An event is composed of a *guard* and an *action*. The guard (written between keywords `where` and `then`) represents conditions that must hold in a state for the event to trigger. The action (written between keywords `then` and `end`) computes new values for state variables, thus performing an observable state transition. If the system reaches a state where no event guard holds, it halts and is said to have *deadlocked*. There is no requirement that an EVENT B system should halt, and indeed, most EVENT B models represent systems that run forever. If halting is desired, the system can be modeled using `convergent` events that monotonically decrease the value of a natural number expression called the machine `variant`. Such events can only be triggered in states where the value of the `variant` is non-negative. Additionally, the system may reach a state where the guards of more than one event hold. In this situation, the system is said to be *non-deterministic*. EVENT B semantics allows any of the events whose guards are satisfied to be triggered.

In EVENT B, systems are typically modeled via a sequence of refinements. First, an abstract machine (model) is written and checked to satisfy a various of invariant properties. Refinement machines are used to add more detail to the abstract machine until the model is sufficiently concrete for hand or automated translation to code. Refinement Proof Obligations (POs) are discharged (proven) to ensure that each refinement is a faithful model of the previous machine so that all machines satisfy the correctness properties of the original.

Table 2.1: **A simplified version of the abstract machine of the chat system**

```
machine machine0 sees ctx0
variables user content chat active

invariants
 @inv1 user ⊆ USER
 @inv2 content ⊆ CONTENT
 @inv3 chat ∈ user ↔ user // chat sessions
 @inv4 active ∈ user ⇸ user // active chat session

events
 event initialization
  then
   @init1 user ≔ ∅   @init2 content ≔ ∅
   @init3 chat ≔ ∅   @init4 active ≔ ∅
 end

 event create-chat-session
 any u1 u2
 where
  @grd1 u1∈user ∧ u2∈user
  @grd2 u1↦u2 ∉ chat
 then
  @act1 chat ≔ chat ∪ {u1↦u2}
  @act2 active(u1) ≔ u2
 end
end
```

Table 2.1 presents a simplified version of an EVENT B model of the chat system. The four invariants typeset the four machine variables. The machine variables store the users and the contents currently in the network, chat sessions, and a variable that keeps track of which chat sessions are currently active, respectively. A chat session for a user `u1` is active if `active(u1)` exists. Since `active` is a function (the symbol ⇸ in @inv4 is used for *partial functions*), if that element exists, then it is unique. On the other hand, `chat` is just a relation (the symbol ↔) so it should be applied to (evaluated over) a set and produces a set of elements. We use square brackets rather than parentheses for applying a relation, hence, `chat[{u1}]` returns the set of elements with whom any member of the singleton set `{u1}` is chatting with.

The `initialization` event gives initial values to the state variables (also called machine variables). The `create-chat-session` event is triggered when any user wants to create a chat session between two users `u1` and `u2`. Guard `@grd2` forbids `create-chat-session` from adding an existing chat session. Action `@act1` modifies the chat to contain the pair of elements `u1↦u2`[1], for which we could have used the more standard notation (`u1, u2`).

Action `@act1` makes the chat `u1 ↦ u2` active for `u1`.

The construct "`any` x `where` G(s,c,v,x) `then` v := A(s,c,v,x) `end`" specifies a non-deterministic event that can be triggered in a state where the event guard G(s, c, v, x) holds for some bounded value x, sets s, constants c, and machine variables v. When the event is triggered, a value for x satisfying guard G(s, c, v, x) is non-deterministically chosen and the event action v := A(s, c, v, x) is executed with x bound to that value. For any x chosen, the new values of the state variables computed by the action of the event maintain the invariant properties of the machine. The semantics of events thus models a system that is controlled by interactions from the environment (i.e., user actions) that may occur at any time.

The example in Table 2.1 uses the Rodin tool (Abrial et al., 2010) notation, where predicates on different lines are implicitly conjoined and actions on different lines are executed simultaneously. EVENT B includes logical notations for conjunction ($\wedge$), disjunction ($\vee$), negation ($\neg$), and implication ($\Rightarrow$). It also includes notation for universal ($\forall$) and existential ($\exists$) quantification.

2.1   RELATIONS AND FUNCTIONS

Functions and relations are in the core of EVENT B's language. EVENT B provides some predefined sets, for instance, the empty set $\varnothing$, the set of natural numbers $\mathbb{N}$, the set of positive numbers $\mathbb{N}1$, and the set of integers $\mathbb{Z}$. It also provides common operations overs sets such as $\cap$, $\cup$, and $\setminus$ for intersection, union, and set difference, respectively. The symbol $\mathbb{P}$ is used for the *power-set* of a set, and the symbol $\times$ is used for the *cross product* between two sets.

Any EVENT B relation is encoded as a set of pairs. The first elements of the pairs are part of the *source* or *domain* of the relation, and the second element of each pair is a member of the *target* or *range* of the relation. A function is a relation such that no two distinct pairs contain the same element. In other words, any element of the source is mapped to at most one element of a target.

Let us suppose that f is a relation with domain A and range B, denoted $f: A \leftrightarrow B$. If f is a function defined for *all* values of A, we say that f is a *total* function, and we write $f: A \rightarrow B$. If f is defined for some values of A, we say that f is a *partial* function, and we write $f: A \nrightarrow B$. If f is a function such as no element in the range of f is associated with more than one element in the

---

[1]   The symbol $\mapsto$ is called maplet, which reads "maps to", hence `u1` maps to `u2`.

domain of f, then we say that f is a *one-to-one* or *injective* function, and we write f : A ⤚⤚ B. If f is a function whose range is B, we say that f is an *onto* or *surjective* function, and we write f : A ⤚⤚ B. If f is both one-to-one and onto, we say that f is a `bijection`, and we write f : A ⤚⤚ B.

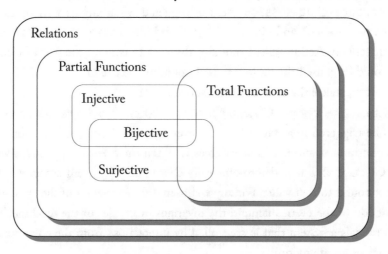

Figure 2.1: Types of mathematical functions.

Figure 2.1 shows a hierarchy of types of relations and functions. Any function is a relation. Functions can be *total* or *partial*. Every total function is a partial function. Functions can be *injective* or *surjective*. Some surjective functions are partial and other ones are total. Some injective functions are partial and other ones are total. Bijective functions are surjective and injective functions. Some bijective functions are partial and other ones are total.

Example 1.   Facebook

Let us suppose that we want to identify the most basic logical structures of a model for Facebook (sets, relations, functions). We are interested in modeling users, social network content, user's pages, content ownership, and permissions on content. We want to model two types of permissions, namely, *view* and *edit* permissions. Pages can be modeled as a relationship from persons to contents: page ∈ person ↔ content. This declaration states that page is an element of all the possible set of relations that can be formed of elements of persons in the domain and network content in the range of the relation.

A relation is just a set of pairs and thus a user page can be defined as page ≙ {(John, Photo1), (John, Video1), (Mary, Photo1), (Mary, Video2)} whereby John has two content items, Photo1 and Video1, Mary has the same photo as John, but has a different video on her page. The domain of page is dom(page) ≙ {John, Mary}, and its range is ran(page) ≙

{Photo1, Video1, Video2}. Notice the use of the symbol "≙" for "is defined as" or "is written as", which should not be confused with the symbol "=", a Boolean operator used for "equals to".

We can enforce a stricter definition for user pages by enforcing that every user has (at least) one content. This is achieved by either asserting the invariant person = dom(page) or, equivalently, by making page a total relation page ∈ person ↔ content. Likewise, one can enforce the property that any content is in someone's page by asserting the invariant content = ran (page) or, equivalently, by declaring page as a surjective relation page ∈ person ↔ content. Finally, we can enforce both properties by declaring page as a total surjective relation page ∈ person ↔ content.

In a social network some users can own certain content, for instance, if a user uploads some content to their page, then the user becomes the owner of the said content. We can model content ownership by introducing a total function from content to users as owner ∈ content → person. The fact that owner is *total* ensures that every content has an owner, and the fact that it is a *function* ensures that the owner is unique. If we further make owner surjective then we force every person to own some (at least one) content: owner ∈ content ↠ person.

We can then define a hierarchy of permission over content, the viewp permission indicates that a user can view or read some content, and the editp permission that she can modify it.

    viewp ∈ content ↠ person
    editp ∈ content ↠ person

Example 2.   View and Edit Permissions

How can we model the property "Users that can edit data (content) must also be able to view it"? We can state the following invariant property:

    editp ⊆ viewp

Notice that we can safely use the ⊆ set operator since a relation is just a set of pairs.

Example 3.   The Content Owner

Model the following property: "The owner of some content has all the permissions on it".

    owner ⊆ viewp
    owner ⊆ editp

Example 4.   Operations Over Sets and Relations

EVENT B is equipped with a series of operators over relations that restrict elements of the relation (pairs) to contain elements in a certain set. Operators ◁, ◁, ▷, and ▹ are used for domain restriction, domain subtraction, range restriction, and range subtraction, respectively. They receive two operands, namely, a relation and a set, whereby the relation filters out some of its elements based on the elements of the set. For instance, the expression r▹s returns a relation r′ after removing all pairs of elements in the relation r whose second components are in the set s. The expression r▹s itself does not modify r but just builds a new relation.

Building on the previous example of Facebook, expression page ▷ {c} returns a relation that is formed from page by considering only the pairs of elements whose second element is c. Expression dom (page▷{c}) returns the set of persons with content c in their page. Expression page▹{c} returns a relation formed from page that disregards any pair of elements whose second element is c. These operators are formally defined in Table 2.2.

Example 5.   Publishing Content on a Social Network

How would you publish a new picture newc in the page of some person p? Publishing content is an operation that typically happens when person p publishes or uploads content into her page. You must modify page, content, owner, viewp, and editp.

```
page  ≔ page ∪ {newc↦p}
content ≔ content ∪ {newc}
owner ≔ owner ∪ {newc↦p}
viewp ≔ viewp ∪ {newc↦p}
editp ≔ editp ∪ {newc↦p}
```

2.2   EVENT B MATHEMATICAL NOTATION

EVENT B provides a full battery of set and relation notation. Table 2.2 shows some of the EVENT B mathematical notations that will be used in this book. We use square brackets to apply (evaluate) a relation to (over) a set of elements as mentioned above. For instance, r[s] applies relation r to all the elements in set s. The result of r[s] is a set of elements in the range of relation r. EVENT B provides standard notations for set union, intersection, difference, etc. The symbol × denotes the cross product between two sets. The operator dom returns the domain of a relation, and ran its range relation. The operator id denotes the identity relation over a set of elements s. Applying the forward composition relation q;r to an element a in the domain of relation q returns a set of elements calculated as the result of applying r to q[{a}]. When relation q is a function, q[{a}] should be used as q(a). The domain restriction relation expression s▷r restricts the domain of a relation r to (consider only elements in) a subset s of its domain. The range restriction relation expression r◁s restricts the range of relation r to consider only elements in a subset s of its range. Domain (range) subtraction is defined similarly to domain (range) restriction, except that the elements in the set s are disregarded rather than considered. EVENT B also provides the r~ notation for the inverse of a relation r.

Table 2.2: EVENT B's basic mathematical notation

Syntax	Name	Definition	Short Form
q;¬r	forward composition	$\{(x,y) \mid \exists y \bullet (x,y) \in q \land (y,z) \in r\}$	q;r
id[s]	identity relation	$\{(x,y) \mid (x,y) \in s \times s \land x = y\}$	id[s]
s◁r	domain restriction	$\{(x,y) \mid (x,y) \in r \land x \in s\}$	id[s];r
s◀ r	domain subtraction	$\{(x,y) \mid (x,y) \in r \land x \notin s\}$	(dom(r)\s)◁r
r▷s	range restriction	$\{(x,y) \mid (x,y) \in r \land y \in s\}$	r▷s
r ▶ s	range subtraction	$\{(x,y) \mid (x,y) \in r \land y \notin s\}$	r▷(ran(r)\s)
r[s]	relational image	$\{y \mid (x,y) \in r \land x \in s\}$	ran(s◁r)
r⊕q	relational overriding	$\{(x,y) \mid (x,y) \in q \lor ((x,y) \in r\} \land \neg \exists z \bullet (x,z) \in q)\}$	qU(dom(q)◁r)
r~	inverse relation	$\{(x,y) \mid (y,x) \in r\}$	r~

Example 6.   Assignments

EVENT B is equipped with two types of assignments: deterministic and non-deterministic. Deterministic assignments use the ≔ notation, and so, for instance, the expression content ≔ cts assigns cts to content. Only machine variables are allowed to be in the left-hand side of an assignment. EVENT B provides two related non-deterministic assignment operators, namely, :| (Becomes such that), and :∈ (becomes in). The right-hand side of :∈ is a set expression and the right-hand side of :| is a predicate. An :∈ expression can always be reduced to a : | expression. The non-deterministic assignment content :| content' ⊇ content assigns to content some of its supersets. The expression content :∈ {content ∪ cts1, content ∪ cts2} non-deterministically assigns to content either content ∪ cts1 or content ∪ cts2. This expression can alternatively be written as content :| content' = {content ∪ cts1} ∨ content' = {content ∪ cts2}.

Building on the example of the social network in order to delete every occurrence of some content c in every single page of Facebook one can use the expression page := page▷{c}. If we additionally want to replace c by some new content newc, then we can use the expression page := page▷{c} ∪ (dom(page▷{c}) × {newc}). Alternatively, to replace c by newc, we can use the expression page := page▷{c} ∪ (page~[{c}] × {newc}), in which page~ is the inverse relation of page and page~[{c}] returns all the persons with c in their page. The symbol × is used for the cross product between two sets.

Example 7.   Quantifiers

EVENT B provides two types of quantifiers, existential and universal quantifiers, denoted ∀ and ∃, respectively. For instance, the expression ∀c,p • c↦p ∈ owner is true of all pair of elements c↦p in owner, and ∃c,p • c↦p ∈ owner is true of (at least) one element in owner.

Example 8.   Booleans

bool, true, and false are Boolean predicates, and BOOL is an enumerated type inhabited by two constants TRUE and FALSE. One could introduce a variable friend ∈ (person ↔ person) ↔ BOOL and add Mary as a friend of John via friend := friend ∪ {(John ↦ Mary) ↦ TRUE}. One can also make Mary a friend of John in case Sara is a friend of Peter through the assignment friend(John ↦ Mary) := bool(friend(Peter ↦ Sara) = TRUE). Friendship is not necessarily a reflexive relationship. That is, the fact that John ↦ Mary ∈ friend does not necessarily mean that Mary ↦ John ∈ friend.

Example 9.   Relational and Functional Overriding

The overriding r⊕q of a relation r by a relation q returns a relation formed taking all the pairs of q and adding all the pairs of r whose first elements are not in the domain of q. Overriding is particularly important when both r and q are functions as it returns another function that updates all the elements in the domain of q. One can update the owner ow of a content item c through the assignment owner(c):= ow or by using the function overriding expression owner := owner⊕{c↦ow}.

Example 10.   Edit Permissions

Remove all the *edit* permissions on some content item c.

```
editp := {c}◁editp
```

Example 11.   Events

We present below the standard syntax used for events; additionally, Table 2.3 explains what each event syntax construct means. The event guard G(c,v,x) depends on the machine constants c, the machine variables v, and the event parameters x, respectively. If an event guard holds, then the event is enabled, and it thus competes with potentially many other events to execute. The body of an event is composed of potentially many event assignments of the form v ≔ E(c,v), where c is the set of machine constants, v is a subset of the set of machine variables, and E is an expression. We use here the ≔ syntax for deterministic assignments, but it can be very well a series of non-deterministic assignments. The symbol `skip` represents the empty set of actions of an event. It does not affect the machine state. The state remains unchanged after `skip` is executed.

```
event name
any x
where
 @grd G(c,v,x)
then
 @act v ≔ E(c,v,x)
end
```

Table 2.3: The Syntax for events

Element	Definition
name	The name of the event
x	Event parameters, a set of variables
G(c,v,x)	Logical predicate
v	A disjoint (sub-) set of machine variables
E(c,v,x)	Sets of expressions

Example 12.   The Chat System: `add-user`

We present below the definition of the event **add-user** which add a fresh user to the chat system. Event guard @grd1 guarantees that user u is a fresh user. Action @act1 adds u to the sets of users of the system. Action @act2 initializes u's chat-content to ∅ (the empty set).

```
event add-user
any u
```

```
where
  @grd1 u ∈ USER\user
then
  @act1 user ≔ user ∪ {u}
  @act2 chatcontent(u) ≔ ∅
end
```

Example 13.   Abstract States (Sets)

We want to model dictionaries in EVENT B. For the sake of simplicity, let us assume that a word has exactly one sole meaning, and that we are just interested in adding words to the dictionary. Variable word is a set of words, and meaning is a set of word meanings. dictionary is a partial function from word to meaning. Therefore, a word has one meaning maximum, yet it might be the case that a word has no meaning.

```
@inv1 word ⊆ WORD
@inv2 meaning ⊆ MEANING
@inv3 dictionary ∈ word ⇸ meaning
```

We show below how to add a new word and its meaning to the dictionary. EVENT B guard @grd1 checks that the word w is in the set difference WORD \ word; since w cannot be in word, it needs to be a new word taken from WORD. Likewise, guard @grd2 checks that m is a new meaning that is not already in the set of meanings. Event action @act1 adds w to the list of words, @act2 adds m to the list of meanings, and @act3 adds to the dictionary an entry that maps word w into meaning m.

```
event add-word
any w m
where
  @grd1  w ∈  WORD \ word
  @grd2 m ∈ MEANING \ meaning
then
  @act1 word ≔ word ∪ {w}
  @act2 meaning ≔ meaning ∪ {m}
  @act3 dictionary ≔ dictionary ∪ {w ↦ m}
end
```

Example 14.   Data Refinement

Let us suppose that we have an abstract state variable a representing a set of natural numbers.

    a: ℙ ℤ

Suppose that we decide to represent this set in our concrete state space using integer sequences (perhaps implemented as linked lists).

    c: seq<ℤ>

Then, an abstract value of {3,5} could be represented by the sequence <3,5> or the sequence <5,3>.

Example 15.   A Segment of Natural Numbers

EVENT B has a built-in notion for a segment of natural numbers based on the notion of set extension. a..b below equals to the set of the natural number between a and b. This set can be empty.

    a..b ≜ {x | x∈ℕ ∧ a≤x ∧ x≤b}

Example 16.   Concrete State (Sequences)

We make variables word and meaning in Example 13 concrete. We implement them as sequences words and meanings, respectively. *words* is encoded as a total function from the set 1..size into word. Likewise, meanings is encoded as a total function from the same set into meaning. We safely assume the same size for both sequences words and meanings since the same meaning cannot be added twice to the dictionary.

```
// number of words and meanings
@invr10 size > 0

//Gluing invariant for "word"
@invr11 words ∈ (1..size) ⟶ word

//Gluing invariant for "meaning"
@invr12 meanings ∈ (1..size) ⟶ meaning
```

Example 17.   **Gluing Invariant for** `word`

We want to add an invariant condition that relates the abstract state variable `word` defined in Example 13 to its concrete state variable `words` in Example 16. This invariant condition is called a *gluing invariant* in literature and is better explained in Section 3.6. The invariant below is a gluing invariant. It says that the abstract variable `word` is equal to the set of all the elements `words(i)` with index i between 1 and `size`.

```
@invr13 word = {i · i ∈ 1..size | words(i)}
```

Example 18.   **Gluing Invariant for** `dictionary`

```
@invr14 ∀i· i∈1..size ⇒
              dictionary(words(i)) = meanings(i)
```

Example 19.   **Adding a** `word` **to the** `dictionary` **(concrete state).**

Event `add-word` below is a *refinement* version of the event defined in Example 13. It assigns into `words` instead of into `word` and `meanings` instead of into `meaning`.

```
event add-word extends add-word
then
  @actr1 size ≔ size+1
  @actr2 words ≔ words ⊗ {size ↦ w}
  @actr3 meanings ≔ meanings ⊗ {size ↦ m}
end
```

2.3   SOFTWARE DEVELOPMENT WITH EVENT B

Software development with EVENT B follows the *parachute strategy* for software development in which systems are first considered from a very abstract and simple point of view, with broad fundamental observations, and then details are added to it until to complete the full functionality of the system. Software development with EVENT B (Abrial, 2010) starts with the definition of an initial *blueprint* of the system we want to model. This blueprint represents the future system implementation. Blueprints give insight on some but not all the aspects of the future system. A blueprint goes through a series of stages called *refinements* (Abrial and Hallerstede, 2007). A blueprint refinement adds details to the blueprint. Refinements provide a hierarchical organisation of the blueprints. The design of the initial system blueprint and its subsequent refinements are based on the description contained in an existing software requirements document. Each stage of the organization of a blueprint serves a different purpose. At higher (more abstract) levels, blueprints are used to state key system properties. At lower (more concrete) levels, blueprints implement the system behavior. It is crucial that the initial blueprint and its refinements are consistent with each other, and that they are coherent with respect to the system specification. A refinement step generates POs. A PO is a proof that the blueprint refinement is a refinement of the blueprint. POs guarantee that the blueprint and its refinement are models of the same system. By discharging (proving) POs one enforces a Correct-by-Construction (CbC) discipline of software development (Hall, 2002; Hall and Chapman, 2002).

EVENT B implements two types of blueprint refinements, *horizontal refinement* (described above) and *vertical refinement* (Abrial, 2009). Horizontal refinement is also called superposition in literature. Horizontal refinements add state transitions to the system or enrich existing transitions. The horizontal refinement stage is complete when all the software requirements are considered in the model. Through horizontal refinement a blueprint can:

- strengthen an event guard,

- add new event guards,

- add more actions to some events, or

- add more events.

Vertical refinement is *data refinement*. It does not add more details to the system, but it transforms the model into something that can easily be implemented. For instance, vertical refinement can transform finite sets into Boolean arrays. We present a concrete example of vertical refinement in Section 3.6 whereby sets are transformed into sequences.

A key aspect of a vertical refinement is the definition of a *gluing invariant*[2] that bridges the abstract state of the system to the concrete state of the system by stating properties of the combined behavior of both state models. Although horizontal and vertical refinements can be combined together in a single refinement step, a final vertical refinement single step is typically realized with the aid of a code generation tool (Cataño and Rivera, 2016; Cataño et al., 2017).

We enumerate below the steps of a software development methodology based on EVENT B. In EVENT B, software development ends up with a program that is correct-by-construction, module the soundness of a code generator that might be used after the definition of the last model refinement, that is, a program that fulfils all the invariant properties of the most abstract EVENT B blueprint.

---

[2] Gluing invariants are discussed in Section 3.6.

1. An initial blueprint (machine) of the system is written in EVENT B. The initial blueprint plays the role of the most abstract view of the system that we want to model.

2. Soundness proofs that demonstrate that each event in the initial blueprint adheres to the invariant properties are conducted, for instance, by using the Rodin platform.

3. A model refinement for the blueprint is written. The refinement can be a horizontal or a vertical refinement, or can combine both.

4. Consistency proofs that demonstrate that the blueprint refinement is a correct refinement of the blueprint are conducted.

5. The two previous steps are iterated as many times as desired.

6. Code is implemented or tool generated for the final model refinement.

The definition of the most abstract machine above and all its refinements are based on an existing software requirements document. The Rodin tool provides support for EVENT B and EVENT B model refinement definition (Abrial et al., 2010). Rodin generates POs in each refinement stage. Rodin comes with various semi-automatic theorem provers that assist in the process of proof discharging.

2.4   A METHODOLOGY FOR EARLY VALIDATION OF SOFTWARE REQUIREMENTS

Early validation of software requirements is of paramount importance in Software Engineering as it minimizes the number of software design errors, hence, preventing the implementation from repeating the same errors. Formal specifications are key to the verification of software requirements. Formal specifications can be used in at least two ways in software development (Almeida et al., 2011). They can be used for model validation, that is, to check if the specification possesses a particular behavior. Or, they can be used for verification purposes, that is, to check or to enforce a formal relationship between the program implementation and its specification. During the model validation process, one can, for example, transform a specification into an executable specification or into a program that can be executed later on. This is often called (specification) *animation*; it is achieved by transforming the specification into an executable format, executing it, and then checking its behavior against some expected behavior. The transformation of a specification into a program can be performed manually or through the use of a tool, for example, using a code-generator. Or, it can combine both approaches, for instance, the formal specification is successively refined (Abrial and Hallerstede, 2007) into a more detailed specification to be finally ported into code, for instance, by using a code generator (Cataño and Rivera, 2016; Cataño et al., 2017). For software requirement validation purposes, the code generated is not meant to be a final software implementation, but merely a prototype that one unit-tests for' flaws. Hence, if some unit-test fails, then the informal and the formal specifications of the system are revisited and evolved.

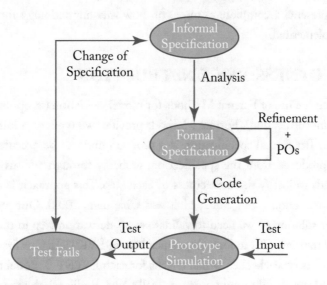

Figure 2.2: Validation approach of software requirements.

The methodology described above follows the model validation approach that is presented in Figure 2.2. In short: (a) we write a software requirements document; (b) we manually port them to formal specifications; (c) we refine the formal specifications as desired; (d) we automatically generate a prototype implementation for the formal specifications; (e) and we finally unit-test the prototype for flaws. If some unit-test fails, then informal and the formal specifications are revisited and evolved.

More concretely, we write a software requirements document in natural language (the informal specification) that includes functional requirements written as User Stories (US), which are textual templates of the form "as a `<Role>`, I want to `<Goal/Desire>` so as to `<Why>`." Although USs are typical of Agile methodologies, we use them here as their syntax closely follows the syntax of the formal language to which we are going to port them (EVENT B). A US includes some Acceptance criteria against which it should be validated. An acceptance criterion has three components, a `Given` part that describes when the functionality may be triggered/executed (which depends on the internal state of the system), a `When` part that tells us when the functionality is to be executed (which depends on the user's decision), and a `Then` part that tells how the state of the system changes when the functionality executes.

After having written all the USs, we manually port them to formal specifications written in EVENT B. Therefore, the `Given` part is formalized as an event guard, the `When` part is the event itself that is executed, and the `Then` part is formalized as a set of event actions.

Section 3.9 presents a complete example on how this methodology for early validation of requirements is implemented.

2.5   CORRECTNESS-BY-CONSTRUCTION

Figure 2.2 focuses on the use of Formal Methods for model validation (as opposed to using Formal Methods for program verification). Formal Methods provide two types of solutions to the problem of model validation. The formal specification is a program and can be executed directly, or some implementation is produced from the specification, probably through various derivation steps, in which case one needs to justify the correctness of each step. This approach is often referred to as Correctness-by-Construction (Hall, 2002; Hall and Chapman, 2002). Our model validation approach fits the latter solution so we need to validate every derivation step in our approach.

We manually translate USs to formal specifications in EVENT B. Our first validation check is carried out with Rodin's type-checker, which checks for each EVENT B model that all its variables, constants, events, and the rest of its constructs are well typed. Rodin raises an error in case it cannot infer the type of any construct. The second validation check is achieved through the generation of POs that attest to the correctness of the EVENT B model. Rodin generates several types of POs (Cataño and Rivera, 2012). It generates a feasibility PO for each action of every event stating that a solution (a program) for the assignment exists. It also generates Os that relates a blueprint and its refinement. For instance, simulation POs ensure that abstract event actions are correctly simulated by concrete actions. Rodin provers are semi-automatic in that they only assist users in discharging the POs.

The last validation check relies on the soundness of the code generator. The translation from EVENT B to JAVA (or any other implementation language) must be sound in order for the JAVA program to attest to the formal specification written in EVENT B. A formal soundness proof for the code generation performed by EVENTB2JAVA has already been conducted (Cataño and Nishi, n.d.).

2.6   RODIN

Rodin is an Eclipse-based platform that provides support to EVENT B, e.g., for writing EVENT B models, defining safety invariant properties, and for discharging POs using back-end provers.[3]

---

[3]   A "prover" is a tool similar to a "proof-assistant"; typically, proof assistants give feedback on the proof process and can backtrack to a previous state of the proof, whereas provers just try to apply a tactic that will eventually succeed or fail.

Rodin is available at https://sourceforge.net/projects/rodin-b-sharp/. The standard way of installing plug-ins in Rodin is through the menu Help, *Install New Software...*, *Add...*, and then adding an Update Site. To conduct proofs in Rodin, one must install the Atelier B provers (the Update Site is http://methode-b.com/update_site/atelierb_provers). It is strongly recommended to use Camille to write and edit EVENT B machines (the Update Site is http://www.stups.hhu.de/camille_updates).

2.7   JML

JML is an interface specification language for JAVA. It is designed for specifying the behavior of JAVA classes, and is included directly in JAVA source files using special comment markers `//@` and `/*@ */`. JML expressions are a superset of Java expressions, with the addition of notations such as `==>` for logical implication, `\exists` for existential quantification, and `\forall` for universal quantification.

JML class specifications can include invariant clauses (assertions that must be satisfied in every visible state of the class), `initially` clauses (specifying conditions that the post-state of every class constructor must satisfy), and history constraints (specified with the keyword constraint, that are similar to invariants, with the additional ability to relate pre- and post-states of a method). Concrete JML specifications can be written directly over fields of the JAVA class.

JML provides pre-post style specifications for JAVA methods describing software contracts. JML uses keywords `requires` for method pre-conditions, `ensures` for normal method post-conditions, and `assignable` and `modifies` for frame conditions (lists of locations whose values may change from the pre-state to the post-state of a method). The pre-state is the state on method entry and the post-state is the state on method exit. A `normal behavior` method specification states that if the method pre-condition holds in the pre-state of the method, then it will always terminate in a normal state, and the post-condition will hold in this state. In a JML `ensures` clause, the keyword `\old` is used to indicate expressions that must be evaluated in the pre-state of the method—all other expressions are evaluated in the post-state. The `\old` keyword can also be used in history constraints, providing a convenient way to specify (for example) that the post-state value of a field is always equal to the pre-state value, thus making the field a constant.

2.8   EVENTB2JAVA

EVENT B specifications are written in the Rodin toolset, an open-source Eclipse IDE that provides a set of tools for working with EVENT B: an editor, a proof generator, and several provers (Abrial et al., 2010). Various code generators for EVENT B exist that take an EVENT B blueprint and generate code in mainstream programming languages such as C or JAVA.

We use EVENTB2JAVA to generate code for the EVENT B specifications that we present. In this book, EVENTB2JAVA is a Rodin plugin so EVENT B modeling and code generation are carried out from Rodin. EVENTB2JAVA relies on two major classes modeling the behavior of classes and relations in JAVA, namely, BSet and BRelation. Class BSet relies on the functionality provided by the JMLEqualsSet class of JML to implement its own functionality. JMLEqualsSet uses a method equals to compare elements in the set and does not clone elements that are added to it. BSet uses a field elems of type JMLEqualsSet to query and change the state of the set. Constant EMPTY is used to represent the ø of EVENT B. Method insert adds an element to the set and method has checks if an element already exists in the set or not.

```
public class BSet<E> extends JMLEqualsSet<E> {
protected JMLEqualsSet<E> elems;

public BSet() {
  elems = new JMLEqualsSet<E>();
}

public static BSet EMPTY = new BSet();

public static <F> BSet<F> singleton(F elem) {
  return new BSet<F>().insert(elem);
}

public BSet<E> insert(E elem) {
  JMLEqualsSet<E> newElems = elems.insert(elem);
  return new BSet<E>(newElems);
}

public boolean has(Object obj) {
  return elems.has(obj);
}

// the rest of the methods

}
```

Class `BRelation` is implemented as a `BSet<Pair<K, V>>`. This follows EVENT B's semantics whereby relations are set of pairs. `BRelation` includes a constant `EMPTY` that encodes an empty relation.

```
public class BRelation<K, V> extends BSet<Pair<K, V>> {
  public static BRelation EMPTY = new BRelation();

  public boolean add(Pair<K,V> pair) {
   return super.add(pair);
  }

  // the rest of the methods

}
```

2.9   A CHAT APPLICATION

We show here some parts of an example that we fully develop in Chapter 3. The example is about a chat application similar to WhatsApp. We use it here to showcase our ideas. User Story US-01 presents an informal requirement that describes the functionality used for creating a chat session between "Me" and "You." The chat may not yet exist.

US-01	create-chat-session
create-chat-session	As a user, I want to create a chat session so as to communicate with You
Acceptance criterion	Given: A chat session between Me and You does not exist When: I decide to create a chat session with You Then: A chat session between Me and You is created, but not between You and Me

US-01 is encoded formally in EVENT B as the event `create-chat- session` below. The event creates a chat session for user Me to chat with user You. The symbol $\in$ is the set membership operation (is a member of), user is the set of users who have signed-up to the chat application. The symbol $\wedge$ is the logical conjunction operator. $\rhd$ is the range restriction operation. Expression {Me} $\rhd$ `active` returns a relation formed from a pair of elements in active whose first component belongs to the singleton set {Me}. Me $\mapsto$ You is a pair of elements, and chat is a mathematical relation that keeps track of the chat sessions currently available. Event guard @grd2 formalizes the Given condition shown in US-01. @act1 encodes the Then expression; it adds the pair of elements Me $\mapsto$ You to the set of existing chats. Guard @grd1 is a typing condition, it helps Rodin's type-checkers infer the type of Me and You.

```
event create-chat-session // US-01
any Me You
where
  @grd1 Me ∈ user ∧ You ∈ user
  @grd2 Me ↦ You ∉ chat
then
  @act1 chat := chat ∪ {Me ↦ You}
end
```

`create-chat-session` can be refined as many times as necessary. We refine it horizontally below (the clause `extends` in EVENT B) by adding more observations or details about creating chat sessions. Our refinement specification below observes active and inactive chat-sessions. `active` is a function that keeps track of active chat sessions. It is a function because only one chat session per user can be active, in other words, a user can actively be chatting to not more than one user. `inactive` keeps track of inactive chat sessions. Action @actr12 makes the chat session added by @act1 in the refined event active. Concepts such as active or inactive are not part of the most abstract formal specification of the chat application, but rather of its first refinement.

```
event create-chat-session extends create-chat-session // US-01
then
  @actr11 inactive ≔ inactive ∪ ({u1} ◁ active)
  @actr12 active(Me) ≔ You
end
```

We show below the example of a unit test that partially validates US-01. Class Pair is a utility class. It encodes pairs of elements of any type. Me and You are integer variables defined globally as class attributes, initialized with values 1 and 2, respectively. Hence, a is the pair Me ↦ You and b is the pair You ↦ Me. The first assertTrue instruction checks that add-user correctly adds Me to user. The second one checks that You is added to user. The third one checks that the pair a is correctly added to chat. Following the Then part of US-01, the last assert instruction checks that b is not added to variable chat.

```
@Test
public void testCreateChat() {
  @act1 chat ≔ chat ∪ {u1↦u2}
  @act2 active(u1) ≔ u2
  add-user(Me);
  add-user(You);
  create-chat-session(Me,You);
  Pair a = new Pair(1,2);
  Pair b = new Pair(2,1);
  assertTrue(user.has(Me));
  assertTrue(user.has(You));
  assertTrue(chat.has(a));
  assertFalse(chat.has(b));
}
```

CHAPTER 3

Software Development of a Chat System with EVENT B

We wrote software requirements for a chat application that includes some of the functionality typically found in WhatsApp (https://www.whatsapp.com/android/). When writing a model for a software system in EVENT B one should write an abstract machine and then successively write refinement machines. For each refinement machine POs are to be discharged in the Rodin platform to ensure that each machine is a proper refinement of the most abstract machines. Only once all the machines are written and all the POs are discharged one can consider the underlying system has completely been modeled. Chapter 4 discusses techniques for discharging POs.

We elicited the requirements for the chat application from our own experience using (the SmartPhone version of) WhatsApp. User Stories relate to the structure of the EVENT B model and the hierarchy of its machines[4]. Table 3.1 presents that hierarchy. The abstract machine (MachineZero) observes basic functionality for chat sessions. The first machine refinement includes functionality to check whether chat content (text, video, picture) can be read or not. The second machine refinement adds implementation details. For instance, it represents content as a sequence (rather than a set) of content items. This is important for us because the graphical interface of a chat session can be implemented as an ordered sequence of content items that reads from the beginning to the end. Additionally, it would help us state a property that says that for any chat session, the chat content reads the same as seen by the two chat members. This is an invariant property that we model here but it is not a property of the SmartPhone version of WhatsApp.

Table 3.1: Machines hieraarchy for chat application

Machine	Observations
MachineZero	The abstract machine, the basic functionality for chat sessions
MachineOne	First refinement, it introduces read and unread status
MachineTwo	Second refinement, it encodes a vertical refinement

Table 3.1 presents the hierarchy of machines of the chat system. The abstract machine observes basic functionality for chat sessions including the functionality for creating a chat session, selecting or un-selecting a chat, chatting, deleting content (text, video, photos), removing content, deleting a chat session, muting and un-muting a chat, and broadcasting and forwarding network content. The first machine refinement includes functionality to check whether chat content has

---
[4] Machine is the name given to a blueprint in EVENT B.

been read or not. The second machine refinement adds implementation details, for instance, it represents content as a sequence (rather than a set) of content items. This is important for us because the graphical interface of a chat session should be implemented as an ordered sequence of content items that reads from the beginning to the end. Additionally, it would help us state a property that says that for any chat session the chat content as seen by one of the two chat members reads the same as it is seen by the other chat member. This is an invariant property. This invariant property, in particular, is not a property of the SmartPhone version of WhatsApp.

Section 3.1 presents the software requirements for MachineZero in Table 3.1, Section 3.2 the software requirements for MachineOne, and Section 3.3 for MachineTwo, respectively. Sections 3.4, 3.5, and 3.6 present the respective formalisation of these software requirements in EVENT B.

3.1   MACHINEZERO

The abstract machine MachineZero contains 10 USs, from US-01–US-10.

US-01 describes the functionality for creating a chat session between "Me" and "You." The chat may not exist already.

US-01	create-chat-session
create-chat-session	As a user, I want to select a chat session so that I can start chatting with You
Acceptance criterion	Given: A chat session between Me and You does not exist When: I decide to create a chat session with You Then: A chat session between Me and You is created, but not between You and Me

US-02 describes the functionality for selecting a chat session. The effect of having two Given conditions is the condition obtained as the conjunction of both.

US-02	select-chat
select-chat	As a user, I want to create a chat session so as to communicate with You
Acceptance criterion	Given: A chat session between Me and You exists Given: A chat session between Me and You is not active When: I select my chat with You Then: The chat session between Me and You is made active

US-03 describes the chatting functionality whereby Me chats with You. Sent content is made available to both of the users Me and You.

US-03	chatting
chatting	As a user, I want to send You some content during a chat session with You to express my ideas
Acceptance criterion	Given: The chat session between Me and You is active When: I chat with You Then: The chat content is made available to Me as well as to You

The chat application implements two different behaviors for deleting exchanged content: "Remove Content for Me" and "Remove Content for Everyone." If the content sender wants to delete some exchanged content, she can remove it from his chat or the chat of the content recipient. On the other hand, the content recipient can only delete the content from her chat. These two behaviors are described by US-04a and US-04b, respectively. Erasing content is the type of subtle functionality that is always difficult to encode in logic as one can easily break invariants. For instance, if one deletes content from one side of the chat and not from the other side, one would break any invariant enforcing that both sides of any should read the same.

US-04a	delete-content
delete-content	As a user, I want to delete some content exchanged with You during a chat session so that I unclutter my chat
Acceptance criterion	Given: The content exists When: Me decides to delete the content that she has received Then: Me's content is deleted

US-04b	remove-content
remove-content	As a user, I want to remove some content exchanged with You during a chat session so that I unclutter my chat
Acceptance criterion	Given: The content exists When: Me decides to remove the content that she has sent Then: The content is removed from Me and anyone to whom Me has sent the content

Chat sessions and associated content can be deleted as well. What US-05 does not state is what would happen with the content seen by You if the session between Me and You is deleted. Will the content be deleted from You as well? Deleting a chat session between Me and You does not delete the content as seen by You, regardless of who sent the content to whom, however, a re-move-content US exists that deletes the content both sides.

US-05	delete-chat-session
delete-chat-session	As a user, I want to delete the chat session between Me and You
Acceptance criterion	Given: A chat session between Me and You exists When: I select to delete the only active chat session Then: The chat session is deleted as well as its associated content

When a chat session has been muted, communication between the two chat users is disabled both ways. Nevertheless, communication can be enabled later on. This is described by US-06.

US-06	mute-chat
mute-chat	As a user, I want to mute a chat session so that I can prevent communication between Me and You
Acceptance criterion	Given: A chat session between Me and You exists When: I decide to mute the chat session between Me and You Then: The chat session is muted and no communication from Me to the muted user or vice-versa is permitted

US-07 is about to re-establish communication between two users of a muted chat. Only the user who muted the chat can unmute it.

US-07	unmute-chat
unmute-chat	As a user, I want to unmute the chat session between Me and You so that I can re-establish the communication with You
Acceptance criterion	Given: The chat session between Me and Another is muted Given: Me muted the chat session between Me and You When: I select to unmute a chat session Then: The communication between Me and You is re-established, that is, the chat between Me and You is unmuted

US-08 and US-09 describe the situation whereby some content is sent to a group of users; forwarding a content requires that respective chat sessions between Me and the group of users exist, broadcasting creates new chat sessions if they do not exist already.

US-08	broadcast
broadcast	As a user, I want to broadcast some content to a group of users so that I can communicate with all of them quickly
Acceptance criterion	When: Me decides to broadcast some content to a group of users Then: The content is sent to all the users in the group

US-09	forward
forward	As a user, I want to forward some content to a group of users so that I can communicate with all of them quickly
Acceptance criterion	Given: Respective chats between Me and Other-Users exist When: Me decides to forward some content to a group of users Then: The content is sent to all the users in the group

US-10 is the counterpart of US-02; unselecting a chat requires the chat to be active.

US-10	unselect-chat
unselect-chat	As a user, I want to unselect the chat session between Me and You so that I can chat with another user
Acceptance criterion	Given: The chat session between Me and You exists Given: The chat session between Me and You is active When: Me decides to unselect her chat with You Then: The chat session between Me and You becomes inactive

Notice that select-chat and unselect-chat could have been written without requiring the chat to be inactive or active, respectively. Thinking about their final encoding, the two events can eventually be encoded by adding a respective checking if-condition that does nothing in case the condition is not fulfilled. On the contrary, by imposing those Given conditions in the US and eventually in their respective EVENT B models we adopt a *defensive* style of modeling in which the system is required to be at the right state to be able to transition to another state.

When modeling a system in EVENT B in addition to the machine's core functionality, one should write a series of invariants that restrict and determine the behavior of the system.

Table 3.2: Invariants for MachineZero	
#	**Invariant**
1	Users are uniquely identified throughout the system
2	Content is uniquely identified throughout the whole system
3	Chat sessions are uniquely identified throughout the system.
4	A chat session relates exactly two users
5	Only one chat session maximum can be established between two users
6	A chat session between two users may have a set of associated content available to either or both of them
7	Content is associated with a chat session only if one the users of the session has sent the content to the other user or vice versa
8	Active and muted are disjoint concepts. That is, it is never the case that the same system reaches a state in which user A muted user B and either is actively chatting with the other one
9	Chat sessions are not symmetric necessarily. That is, the fact that user A has created a chat session to chat with user B, does not necessarily mean that user B has a created session to chat with user A
10	Active chat sessions are not symmetric necessarily. That is, the fact that user A is actively chatting with user B does not necessarily mean that user B is actively chatting with user A
11	It is never the case that chat content exists associated with a pair of users for which no chat session exists
12	Several chat sessions can be created, but only one (or none) created chat session may be active per user
13	Chat communication with a muted user is no feasible: no content exchange is feasible from or to a muted chat

3.2   MACHINEONE

Our chat system resembles WhatsApp, which implements a graphical interface with one grey tick indicating that the message has successfully been sent to the recipient's device, and two grey ticks indicating that the recipient has read the message. EX-01 describes the latter functionality. A content item is read if the recipient has an active chat session with the sender during or after the content item has been sent. In other words, once a user opens up a window chat, the chat content is considered to have been read.

EX-01	Read Stamp
Read Stamp	As a user, I want to know I have already read my messages with You so as to keep track of my pending messages
Acceptance criterion	Given: A chat session between Me and You exists When: Any time Then: A Read/Unread stamp for the chat session between Me and You produced

Table 3.3: Invariants for MachineOne

#	Invariant
1	Every chat session has an associated "read" or "unread" status.
2	A chat session has a "read" status if each piece of content that it contains has a "read" status. If some chat content has an "unread" status associated then the complete chat session has an "unread" status associated

3.3   MACHINETWO

EX-02 offers a general description of the functionality for reading a chat session. Reading a chat requires the chat content to be ordered some way.

EX-02	reading-chat
reading-chat	As a user, I want to read the content of the chat session between Me and You
Acceptance criterion	Given: The chat session between Me and You exists When: I decide to read the chat session between Me and You Then: The content of the chat session between Me and You is made available to Me

3.4   MACHINEZERO IN EVENT B

We present here the formalisation of MachineZero. A key concept is machine invariant. Invariants are of various kinds in EVENT B, namely: (i) typing invariants such as user $\in$ USER, which states that user is a subset of USER; (ii) gluing invariants, which relate the abstract state of a blueprint with its concrete state; and (iii) invariants that restrict the state of variables. Only invariants of the third kind above are likely to be included in a software requirements document. Invariants falling in the first and second invariants category are not included since they are particular to EVENT B.

We start by looking at the context of the abstract machine, which introduces two *carrier sets*, namely, USER and CONTENT that are used to typeset all the users registered in the chat system and all the content that it manipulates.

```
context ctx0
 sets USER CONTENT
end


machine machine0 sees ctx0
 variables user content chat active chatcontent muted

// machine invariants...

  event initialization extends initialization
   then
     @init1 user ≔ ∅      @init2 content ≔ ∅
     @init3 chat ≔ ∅      @init4 active ≔ ∅
     @init5 chatcontent ≔ ∅ @init6 muted ≔ ∅
   end


   // machine events...


  end
```

The first two invariants in Table 3.2 are implemented by introducing two set variables in our model; the first one holds the users who have signed up to the chat system, and the second one holds all the exchanged content.

```
invariants
  @inv1 user ⊆ USER
  @inv2 content ⊆ CONTENT
```

Invariant 3 in Table 3.2 is implemented as an invariant that declares chat as a relationship between users. Invariant 4 is implemented by the fact that chat is a binary relation. Having modeled chat as a set enforces the fifth invariant in Table 3.2, therefore, no pair of elements in a chat session is repeated.

```
@inv3 chat ∈ user ↔ user // chat sessions
```

Implementing the sixth invariant in Table 3.2 requires a subtler analysis as it relates content, the sender and the receiver of the content. Variable `chatcontent` below introduces chat content. The variable is defined as a partial function with domain `user` (the person who sends the message) and range `content` ⇸ ℙ(`user`), where `content` is the content sent and ℙ(`user`) is the set of users to whom the content has been sent. `chatcontent` is a *partial function*, therefore, it might be the case a user exists that has not chatted with anyone. The range of `chatcontent` is again a *partial function*, therefore, it might be the case a user exists that has not chatted with some particular user. Since `chatcontent` and its range are functions, the set of users to whom user `u1` has sent some content `c` is uniquely represented as `chatcontent(u1)(c)`, given that `u1` exists in the domain of `chatcontent` and `c` exists in the domain of `chatcontent(u1)`. The set of users with whom `u1` has chatted is represented as `ran(chatcontent(u1))`, and the set of content items sent by `u1` (to anyone) is represented as `dom(chatcontent(u1))`, given that `u1` exists in the domain of `chat-content`.

```
@inv4 chatcontent ∈ user ⇸ (content ⇸ ℙ(user))
```

Next, we proceed to encode invariant 8 in Table 3.2, which says that active and muted chats are disjoint. `@inv5` below encodes the set of active chat sessions; `active` is a *partial function*, hence, a user has one active chat session maximum (the *function* part), but it might be the case he has no active chat session at all (the *partial* part). `@inv7` states that it is never that case an active chat session is not a chat session, and `@inv8` states that it is never the case that a muted chat session is not a chat session, therefore, elements from muted chats are taken from chats. `@inv9` encodes invariant 8 in Table 3.2.

```
@inv5 active ∈ user ⇸ user // active chat session
@inv6 muted ∈ user ↔ user // muted sessions
@inv7 active ⊆ chat // active chat sessions
@inv8 muted ⊆ chat // muted chat sessions
@inv9 muted ∩ active = ∅
```

Invariants 9 and 10 in Table 3.2 state that `chat` and `active` sessions are not symmetric necessarily. These invariants are modeled by not imposing further constraints over them. In other words, if we want them to be symmetric, we need to enforce further invariants. Invariant 11 in Table 3.2 is implemented by `@inv10` below. Expression `chatcontent(u)(c)` returns the set of users to whom user u has sent some content. Expression `chat[{u}]` returns the set of users with whom user u is chatting.

```
@inv10 ∀u,c·u∈user ∧ c∈ content ∧
u∈dom(chatcontent) ∧ c∈dom(chatcontent(u)) ⇒
                        chatcontent(u)(c) ⊆ chat[{u}]
```

Invariant 12 in Table 3.2 is enforced by the fact that `active` is a function.

EVENT B models are composed of a *static part* defining observations (variables, constants, parameters, etc.) of the system and their invariants properties, and a *dynamic part* defining operations (events) changing the state of the system. Definitions introduced up to now are all static, and the next definitions are the dynamic part of the abstract machine (`machine0`) of our model. Invariant 13 in Table 3.2 is dynamic. It requires us to add an event guard to every event that otherwise might modify `chatcontent` of a muted chat.

Next, we implement the basic functionality of chat sessions in EVENT B.

Event `create-chat-session` implements US-01. It creates a chat session for user u1 to chat with user u2. The `Given` condition in **US-01** is encoded by guard @grd2. Guard @grd1 is a typing condition, it helps Rodin to infer the type of u1 and u2. @act1 adds the pair u1↦u2 to the set of existing chats. @act2 makes the content associated with the chat between u1 and u2 empty. Notice that event `create-chat-session` does not create a chat for u2 to chat with user u1.

```
event create-chat-session // US-01
any u1 u2
where
  @grd1 u1∈user ∧ u2∈user
  @grd2 u1↦u2 ∉ chat
then
  @act1 chat ≔ chat ∪ {u1↦u2}
  @act2 active(u1) ≔ u2
end
```

Event `select-chat` implements US-02. @act1 uses the relational overriding operator ⊗ instead of the set union operator ∪, in this way u1 can have an active chat session only with one user. @grd4 implements a defensive style of programming as explained before. @grd3 makes sure that a muted chat session is never active. @grd1 is a typing guard. It typesets u1 and u2. @grd2 implements the first Given condition in US-02, and guard @grd4 implements the second one. @act1 uses the overriding operator ⊗ instead of the union operator ∪ to make sure we don't break @inv5 so that active remains a function. Had we added u1 ↦ u2 to active using the union operator ∪, we would have probably ended up with active mapping u1 to two different users. Rodin would have detected this mistake by generating an unprovable PO.

```
event select-chat // US-02
any u1 u2
where
 @grd1 u1∈user ∧ u2∈user
 @grd2 u1 ↦ u2 ∈ chat
 @grd3 u1 ↦ u2 ∉ muted
 @grd4 u1 ↦ u2 ∉ active
then
 @act1 active ≔ active ⊗ {u1↦u2}
end
```

Event chatting implements US-03 whereby user u1 chats with user u2. It implements the scenario whereby u1 sends some content c to u2. @grd2 encodes the Given condition. The first part of guard @grd4 typesets variable c and the second part requires it to be some fresh content. Because c is fresh, @act1 adds it to the set of contents. @act2 creates a chat instance for u2↦u1 in case it does not exist already. If it exists, chat remains unchanged since it is a set. This matches the actual behavior of WhatsApp in which a chat window is created for u2 the first time a user u1 sends her some content. The second line in @act3 adds c to the existing chat content between u1 and u2. Notice that chatcontent(u1) remains a function after the assignment in @act3 since c is not in its domain.

```
event chatting // US-03
 any u1 u2 c
 where
  @grd1 u1∈user ∧ u2∈user
  @grd2 u1 ↦ u2 ∈ active
  @grd3 u1 ↦ u2 ∉ muted ∧ u2↦u1 ∉ muted
  @grd4 c ∈ CONTENT ∧ c ∉ content
```

```
  then
    @act1 content ≔ content ∪ {c}
    @act2 chat ≔ chat ∪ {u2↦u1}
    @act3 chatcontent ≔ chatcontent ⊕
                        {u1↦ chatcontent(u1) ∪ {c↦{u2}}}
  end
```

We present below the encoding of US-04a and US-04b; therefore, `delete-content` removes content c from the chat u1↦u2, and `remove-content` from everywhere. Notice that `delete-content` does not remove c from chat u2↦u1. Guard @grd2 verifies that the user u1 who deletes or removes the content is actively chatting with u2. `delete-content` uses the functional overriding operator ⊗ to override u1's chat content. It removes u2 from `chatcontent(u1)(c)` so that u2 no longer appears as having received content c from u1.

`remove-content` uses the domain subtraction operator ◁ to remove content c from every user to whom u1 sent the said content.

Notice that if u1 is chatting with u2, and u2 with u3, and u1 sends c to u2, and u2 sends c to u3, calling `remove-content` with parameters u1, u2, and c does not remove c from the chat between u2 and u3, but only from the chat between u1 and u2 and between u2 and u1. For this very reason, `remove-content` does not implement a second action @act2 content ≔ content \ {c}.

```
  event delete-content // US-04a
    any u1 u2 c
    where
      @grd1 u1∈user ∧ u2∈user
      @grd2 u1↦u2 ∈ active
      @grd3 u1 ∈ dom(chatcontent)
      @grd4 c ∈ dom(chatcontent(u1))
      @grd5 u2 ∈ chatcontent(u1)(c)
    then
      @act1 chatcontent(u1) ≔ chatcontent(u1) ⊗
                            {c ↦ (chatcontent(u1)(c)\{u2})}
  end


  event remove-content // US-04b
    any u1 u2 c
    where
```

```
   @grd1 u1∈user ∧ u2∈user
   @grd2 u1↦u2 ∈ active
   @grd3 u1 ∈ dom(chatcontent)
   @grd4 c ∈ dom(chatcontent(u1))
   @grd5 u2 ∈ chatcontent(u1)(c)
 then
   @act1 chatcontent(u1) ≔ {c} ◁ chatcontent(u1)
   @act2 content ≔ content \ {cc
 end
```

Event `mute-chat` implements US-06. It mutes the chat between u1 and u2; more concretely, `@act1` adds the pair u1 to the set of muted chats. `@act2` forbids a muted chat from being active. Alternatively, we could have added a guard `@grd4 u1 ↦ u2 ∉ active`, but then this does not reflect the actual behavior of the graphical interface of WhatsApp in which a user u1 can indeed mute a user u2 when actively chatting with her.

```
event mute-chat // US-06
 any u1 u2
 where
  @grd1 u1∈user ∧ u2∈user
  @grd2 u1 ↦ u2 ∈ chat
  @grd3 u1 ↦ u2 ∉ muted
 then
  @act1 muted ≔ muted ∪ {u1↦u2}
  @act2 active ≔ active \ {u1↦u2}
 end
```

Event `unmute-chat` implements US-07. It unmutes the chat between u1 and u2. `@grd3` verifies that u1 had muted u2. If it exists, the user who muted u2 is unique since muted is a function. `@grd2` is redundant: it can be deduced from `@grd3` and the fact that `muted ⊆ chat`. Nevertheless, although redundant, `@grd2` helps Rodin type the event guards and helps us and other readers understand our model better.

```
event unmute-chat // US-07
 any u1 u2
 where
```

```
    @grd1 u1∈user ∧ u2∈user
    @grd2 u1 ↦ u2 ∈ chat
    @grd3 u1 ↦ u2 ∈ muted
  then
   @act1 muted ≔ muted \ {u1↦u2}
  end
```

Event `forward` below implements US-09 whereby user u forwards content c to a set of users us. Guards @grd1, @grd2, and @grd5 typeset u and us. @grd4 typesets c. @grd6 checks that u indeed possesses chat sessions with every user member of us. Expression muted[{u}] ∩ us = ∅ checks that no member of the set us is part of the set of users that u has muted. Expression muted[us] ∩ {u} = ∅ verifies that u has not been muted by any member of us. Body action @act2 creates respective chat sessions for each member of the set us to chat with u. Action @act1 overrides chatcontent to include content c into the chat sessions between u and each element of us. Event `forward` itself does not implement a notion of order among the content items of a chat session, hence, at the abstract level as implemented by the abstract machine one cannot establish which chat content reads first or after another.

Event `broadcast` for US-08 is implemented in a similar way to event `forward`. The major difference between both of them is that guard @grd6 for event `forward` is not included by event `broadcast`.

```
event forward // US-09
  any u us c
  where
   @grd1 u ∈ user
   @grd2 us ⊆ user
   @grd3 muted[{u}] ∩ us = ∅ ∧ muted[us] ∩ {u} = ∅
   @grd4 c ∈ content
   @grd5 u ∈ dom(chatcontent)
   @grd6 us ⊆ chat[{u}]
  then
   @act1 chatcontent ≔ chatcontent ⊕ {u↦{c↦us}}
   @act2 chat ≔ chat u (us×{u})
  end
```

Event `unselect-chat` implements US-10. It unselects the chat between `u1` and `u2` by dropping `u1↦u2` from `active`.

```
                              // US-10

event unselect-chat
any u1 u2
where
 @grd1 u1∈user ∧ u2∈user
 @grd2 u1 ↦ u2 ∈ chat
 @grd3 u1 ↦ u2 ∈ active
then
 @act1 active ≔ active \ {u1↦u2}
end
```

3.5   MACHINEONE IN EVENT B

Machine refinement is the mechanism that EVENT B offers to extend or to detail the behavior and functionality of a machine. In EVENT B, all the components of a *refined machine* are included in a *refinement machine*, either explicitly (machine variables, machine variable initializations, guards and actions of a refining event defined using `refines`) or implicitly (invariants, guards, and actions of a refining event defined using `extends`). Machine `machine1` below sees `machine0`, hence it refines all its functionality. It additionally introduces variable `toread` so as to encode US EX-01. This variable stores all the chat-content pending to read.

Variables `user`, `content`, `chat`, `active`, `chatcontent`, and `muted` below are explicitly refined by `machine1`. The machine includes two further variables `toread` and `inactive` which are initially empty. Notice that the `initialization` event does not initialize any of the abstract variables `user`, `content`, `chat`, `active`, `chatcontent`, or `muted`, as they are already initialized in `machine0` (the most abstract machine). Machine `machine1` "sees" the same context `ctx0` that `machine0` sess.

```
machine machine1 refines machine0 sees ctx0

variables user content chat active chatcontent muted
  toread // content pending to read
  inactive // inactive chats

  // refined invariants...
```

```
event initialization extends initialization
 then
  @initr11 toread ≔ ∅
  @initr12 inactive ≔ ∅
 end


// refined events...


end
```

Invariants @invr11 and @invr12 declare `toread` and `inactive` as relations of type `user` ↔ `user` which are included in `chat`. Invariant @invr13 constrains `chat` to be the union of active, `toread`, and `inactive`; active and `toread` are disjoint sets, as well as active and inactive, however, a chat session may be `inactive` and `toread` at the same time.

```
invariants
 @invr11 toread ⊆ chat
 @invr12 inactive ⊆ chat
 @invr13 active ∪ toread ∪ inactive = chat
 @invr14 active ∩ toread = ∅
 @invr15 active ∩ inactive = ∅
```

US EX-01 is implemented *statically* by all the invariants manipulating variables `toread` or `active`, and *dynamically* through the encoding of events within `machine1` that refer to either variable.

Machine events are implicitly refined within the refinement machine through the use of an `extends` clause. The parameters of the refined event `unmute-chat` below are `u1` and `u2` union the new parameters that `unmute-chat` declares in `machine1` (the empty set). The guards of the refined event `unmute-chat` below are @grd1, @grd2, and @grd3 from the abstract event conjoined with the new guards that `unmute-chat` declares in `machine1` (none). The same goes for the abstract action of event `unmute-chat`, that is @act1 `muted ≔ muted \ {u1↦u2}`, which implicitly appears repeated in the concrete version of unmute–chat event below.

```
event unmute-chat extends unmute-chat // US-07
end
```

Next, we describe how local invariants in Table 3.3 are implemented in EVENT B. Regarding Invariant 1, the "unread" status is encoded through the variable `toread`, and the "read" status as `chat \ toread`. Notice that our EVENT B encoding is more granular than what is described by the USs and the local invariants since a variable possessing a "read" status can be `active` or `inactive`. This is often the case in Software Engineering where USs are kept away from implementation details.

Invariant 2 cannot be encoded statically directly through invariants but it rather needs to be encoded through the machine events, hence every time a content item is sent to an `inactive` chat then it needs to be added to `toread`.

We show the event `create-chat-session` below, where the expression `{u1} ◁ active` is equivalent to the singleton set `{active(u1)}` whenever `u1` is in the domain of `active`. More concretely, `active(u1)` requires `u1` to exist in the domain of `active`, whereas `{u1} ◁ active` does not and it hence can sometimes return ø. Therefore, every time some user `u1` creates a chat session, the chat session between her and the user she is actively chatting with (if any) is made `inactive`. Because `active` is a function, whenever `active(u1)` exists, then expression (`{u1} ◁ active`) becomes the singleton set `{u1 ↦ active(u1)}`.

```
event create-chat-session extends create-chat-session // US-01
then
  @actr11 inactive ≔ inactive ∪ ({u1} ◁ active)
end
```

All the content items of a selected chat session are read all at once. This is modeled by `@actr11` in `select-chat`. `active` and `inactive` are two disjoint sets as given by invariant `@invr15`. Because the abstract event `select-chat` adds `u1↦u2` to `active`, we need to drop it from `inactive` to keep invariant `@invr13`. This is modeled by the left disjunction of `@actr12`. If `u1` was actively chatting with some user, then the chat between the former and the latter is made inactive too. This is modeled by the right disjunction of `@actr12`.

```
event select-chat extends select-chat // US-02
then
  @actr11 toread ≔ toread \ {u1↦u2}
  @actr12 inactive ≔ (inactive \ {u1↦u2}) ∪ ({u1} ◁ active)
end
```

The refinement event `chatting` is presented below. When u1 is chatting with u2, which chat sessions should be added to `toread`? It should be added u2↦u1 since a content c has been sent from u1 to u2, except in the case that u2↦u1 is in `active`. This is modeled by the expression ({u2↦u1} \ `active`). We should also add the pair u2↦u1 to `toread` when the pair is `inactive`. This is modeled by the expression ({u2↦u1} ∩ `inactive`). But then, if that is the case, we need to drop out u2↦u1 from `inactive`. The latter is achieved by @actr12.

```
event chatting extends chatting // US-03
then
  @actr11 toread := toread ∪
                      ({u2↦u1} \ active) ∪
                      ({u2↦u1} ∩ inactive)
  @actr12 inactive := inactive \ ({u2↦u1} ∩ inactive)
end
```

Event `delete-content` remains the same. No behavior is added by this refinement event.

```
event delete-content extends delete-content // US-04
end
```

When deleting chat u1↦u2, one should drop it out from the list of inactive or pending-to-read chats. Every pending action for a deleted chat session ceases to exist.

```
event delete-chat-session extends delete-chat-session // US-05

then
  @actr11 toread := toread \ {u1↦u2}
  @actr12 inactive := inactive \ {u1↦u2}
end
```

The refinement event `mute-chat` is presented below. Because the abstract event drops out the pair u1↦u2 from `active`, then the concrete event below adds it to `inactive` so as to preserve invariant @invr13 of `machine1`.

```
event mute-chat extends mute-chat // US-06
then
```

```
  @act11 inactive ≔ inactive ∪ {u1↦u2}
  @act12 toread ≔ toread \ {u1↦u2}
end
```

The implementation of `forward` is similar to the implementation of `chatting` by taking `{u2↦u1}` to be equal to `(us×{u})∪({u}×us)`.

```
event forward extends forward // US-09
then
 @actr11 toread ≔ toread ∪
           (((us×{u}) ∪ ({u}×us)) \ active) ∪
           (((us×{u}) ∪ ({u}×us)) ∩ inactive)
 @actr12 inactive ≔ inactive \ (((us×{u})∪({u}×us)) ∩ inactive)
end
```

The implementation of `broadcast` is exactly the same as the implementation of `forward`.

Unselected chats are made inactive.

```
event unselect-chat extends unselect-chat // US-10
then
 @actr21 inactive ≔ inactive ∪ {u1↦u2}
end
```

3.6   MACHINETWO IN EVENT B

Machine `machine2` adds implementation details to our EVENT B model of WhatsApp2. The goal of this machine is to leave the EVENT B model in a way that is close to implementation for it to be translated to JAVA using the EVENTB2JAVA tool (Cataño and Rivera, 2016). Variables of a refined machine can appear in an invariant of a refinement machine. When this happens, the invariant is called a *gluing invariant* as it relates the state space of the abstract (refined) machine with the state space of the refinement machine. Until now, content, users, and chat content have been modeled as sets. We want to model them as sequences through the use of gluing invariants.

In EVENT B, a segment of natural numbers can be expressed using the a..b notation, which defines the set of natural numbers between a and b inclusive.

```
a..b = {x | x∈N ∧ a≤x ∧ x≤b}
```

We can hence use the 1..n notation to model a sequence of type T and size n as a total function from 1..n to T. By requiring sequence to be a *total function* we enforce it to have no holes in its domain.

```
@inv sequence ∈ 1..n ⟶ T
```

Variable contents encode content as a sequence. csize represents the number of content items in contents. content is the type of contents. The domain of contents is 1..csize, hence, when csize is 0, contents is empty.

```
@invr21 csize ≥ 0
@invr22 contents ∈ (1..csize) ⤖ content
@invr23 content = {n,c · n↦c ∈ contents | c}
```

Next, we choose to implement chatcontent as the variable screen. This refined variable makes content-sent sequential but not the sender or the receiver of the content. This is because we mainly use screen to display content exchanged in an orderly fashion, for which the pair of users do not need to be ordered, just the content items.

```
@invr24 screen ∈ user ⤔ (user ⤔ P(contents))
```

We present the refined version of chatting below; parameter k1 is the position at which content c is placed at u1's screen. For u1's chat content to be shown in an orderly fashion, k1 must be greater than any value in dom(screen(u1)(u2)) every time that chatting executes. Likewise, k2 is the position of content item c in u2's chat screen with u1. The last conjuncts in @grdr21 and @grdr22 together ensure that P(contents) in @invr24 is a function. Action @actr22 increases the number of existing content items. Action @actr23 adds c at position csize+1 of sequence contents. @actr21 adds c at position k1 (k2) of u1's (u2's) chat screen with u2 (u1).

```
event chatting extends chatting
any k1 k2
```

```
where
  @grdr21 u1∈dom(screen) ∧ u2∈dom(screen(u1)) ∧
          k1 ∉ dom(screen(u1)(u2))
  @grdr22 u2∈dom(screen) ∧ u1∈dom(screen(u2)) ∧
          k2 ∉ dom(screen(u2)(u1))
then
  @actr21 screen ≔ screen ⊗
            {u1↦ (screen(u1) ⊗ {u2↦ (screen(u1)(u2) ⊗ {k1↦c})}),
             u2↦ (screen(u2) ⊗ {u1↦ (screen(u2)(u1) ⊗ {k2↦c})})}
  @actr22 csize ≔ csize+1
  @actr23 contents ≔ contents ⊗ {csize+1↦c}
end
```

Event `delete-content` declares two parameters, `i` and `k`, which represent the position of content `c` in the sequence `contents` and the sequence `screen`, respectively. `@grdr21` checks that `contents(i) = c`. The last conjunct of `@grdr22` checks that `c` is displayed at position `k` of the chat screen between `u1` and `u2`. `@actr21` deletes `k↦c` from `screen(u1)(u2)`.

```
event delete-content extends delete-content
any i k
where
  @grdr21 i↦c∈contents
  @grdr22 u1∈dom(screen) ∧ u2∈dom(screen(u1)) ∧
          k↦c∈screen(u1)(u2)
then
  @actr21 screen(u1) ≔ screen(u1) ⊗
                        {u2 ↦ ({k} ⊗ screen(u1)(u2))}
end
```

The remaining events of `machine2` are defined similarly. For instance, `forward` and `broadcast` are defined similar to `chatting`, and `delete-chat-session` similar to `delete-content`.

We present below the concrete version of the `add-user` event as defined in machine2. The abstract version is presented in Example 12. `screen(u)`'s content is made empty as seen by every `user` in the system.

```
event add-user extends add-user
then
 @actr22 screen(u) ≔ user × {∅}
end
```

3.7   THE IMPLEMENTATION OF THE CHAT SYSTEM

EVENTB2JAVA translates an EVENT B machine into a JAVA class. In translating a machine, EVENTB2JAVA not only considers the information provided by the machine, but also the contexts the machine "sees." Refinement machines are translated in the same way as abstract machines since Rodin properly adds abstract machine components to the internal representation of the refining machine. EVENTB2JAVA generates JML specifications in addition to the JAVA code. JML is an interface specification language for JAVA. It is designed to represent the behavior of JAVA classes, and it is included directly in JAVA source files using special comment markers. Our presentation of WhatsApp implementation here focuses on JAVA mainly.

EVENTB2JAVA translates Invariants only as JML specifications. The translation of constants includes a JAVA part and a JML part. EVENT B logical predicates are translated to JAVA predicates written with the aid of classes BSet and BRelation, implementing sets and relations in JAVA, respectively. Refining and extending events are translated in the same manner as abstract events. Each event is translated to a separate JAVA class. The translation of each event includes an object reference to the machine class. The translation of a standard (non-initializing) event includes a guard-event method that tests if the guard of the event holds, and a run-event method that models the execution of event.

We show below the code that EVENTB2JAVA generates for event add-user. Method guard-add-user below implements the guard of event add-user and method run-add-user its actions.

```
public class add-user {
  private machine0 m; // reference to the machine
  public add-user(machine0 m) { this.m = m; }
  public boolean guard-add-user(Integer u) {
    return m.USER.difference(m.get-user()).has(u);
  }
  public void run-add-user(Integer u){
    if(guard-add-user(u)) {
    BSet user = m.get-user();
```

```
      BRelation chatcontent = m.get-chatcontent();
      m.set-user(user.union(new BSet(u)));
      m.set-chatcontent(chatcontent.override(new BRelation(
         new Pair(u,BRelation.cross(m.get-content(),
              new BSet(BRelation.EMPTY))))));
    }
   }
  }
```

The implementation of event `create-chat-session` is shown below. Variables `u1` and `u2` are parameters for the "guard" and "run" class methods. `guard-create-chat-session` is a Boolean method that returns `true` when the event guard holds. `run-create-chat-session`[5] implements two event actions, one modifies `chat` and the other one `active`. Variable `m` is an object instance of class `machine2`. This class intensively uses "get" and "set" methods to access and modify the machine variables.

```
public class create-chat-session { // US-01
  private machine2 m;

  public boolean guard-create-chat-session(Integer u1, Integer u2) {
    return (m.get-user().has(u1) && m.get-user().has(u2) &&
    !m.get-chat().has(new Pair(u1,u2)));
  }

  public void run-create-chat-session(Integer u1, Integer u2) {
    if(guard-create-chat-session(u1,u2)) {
     BRelation chat = m.get-chat();
     BRelation active = m.get-active();

     m.set-chat((chat.union(new BRelation(new Pair(u1,u2)))));
     m.set-active((active.override(new BRelation(new Pair(u1,u2)))));
    }
   }
  }
```

---

[5] `!` is the symbol for logical negation in JAVA.

We show below the implementation of machine2. Constants CONTENT and USER are defined in the machine context ctx. We used them to typeset some machine variables content and user. They have Enumerated type and are declared to contain all the elements between min-int and max-int. Class machine2 includes not only the variables declared in machine machine2 but also machine0.

The generated code heavily uses the functionality provided by classes BSet and BRelation; these classes are briefly discussed in Section 2.8.

```java
public class machine2 {
  private static final Integer max-int = Utilities.max-int;
  private static final Integer min-int = Utilities.min-int;

  public static final BSet CONTENT = new Enumerated(min-int,max-int);
  public static final BSet USER = new Enumerated(min-int,max-int);

  private BRelation active;
  private BRelation chat;
  private BRelation chatcontent;
  private BSet content;
  private BRelation contents;
  private Integer csize;
  private BRelation muted;
  private BRelation screen;
  private BSet user;

  private BRelation get-chatcontent(){
    return this.chatcontent;
  }

  public void set-chatcontent(BRelation chatcontent){
    this.chatcontent = chatcontent;
  }

  // the rest of the set- and get- methods
  // for the rest of variables ...
}
```

3.8   TESTING AND CODE ANIMATION OF THE CHAT SYSTEM

The code generated by the EVENTB2JAVA tool can be animated and tested using JUnit (JAVA Unit) tests (Appel, 2015). Testing allows one to validate if the code implemented in JAVA behaves as expected.

Class `add-contentTest` implements a simple JUnit test for functionality `add-content`. JUnit classes are mainly composed of three methods:

- a `@Before` `setUp()` method that executes when the test class object is created, and before any `@Test` `test()` method;

- an `@After` `tearDown()` method that executes when the test class object is destroyed, after any `@Test` `test()` method can run; and

- one or many `@Test` `test()` methods that test the functionality of the class; they typically use `assertTrue` and `assertFalse` methods to check the expected behavior of the class object. `assertTrue` and `assertFalse` receive a Boolean expression as a parameter.

Method `testAddContent()` tests different aspects of its implementation:

- `assertTrue(contents.size() == 1);` checks that the first time that one adds a content item, the size of contents is 1;

- `assertTrue(contents.has(c1));` checks that the content item added is indeed added; and

- `assertFalse(contents.has(c2));` is a redundant test that can be deduced from the two previous tests.

```
public class add-contentTest {
  Integer c1, c2;
  machine2 machine;

  @Before
  protected void setUp() throws Exception {
    c1 = 1;
    c2 = 2;
    machine = new machine2();
  }
```

```
@After
protected void tearDown() {
  c1 = 0;
  c2 = 0;
  machine = null;
}


@Test
public void testAddContent() {
  machine.add-content(c1);
  BSet<Integer> contents = machine.get-content();

  assertTrue(contents.size() == 1);
  assertTrue(contents.has(c1));
  assertFalse(contents.has(c2));
}


}
```

The JUnit testing add-user is similar to testing add-content:

- assertTrue(users.size() == 1); tests that the number of users after adding u1 is 1;

- assertTrue(users.has(u1)); tests that users contains u1 after it has been added to it; and

- assertFalse(users.has(u2)); is a redundant test that can be inferred from the two previous tests.

```
public class add-userTest {
  Integer u1, u2;
  machine2 machine;

  @Before
  protected void setUp() throws Exception {
    u1 = 1;
```

```
  u2 = 2;
  machine = new machine2();
}


@After
protected void tearDown() {
  u1 = 0;
  u2 = 0;
  machine = null;
}


@Test
public void testAddUser() {
  machine.add-user(u1);
  BSet<Integer> users = machine.get-user();

  assertTrue(users.size() == 1);
  assertTrue(users.has(u1));
  assertTrue(users.has(u2));
}


}
```

Method `testCreateChat()` in class `create-chat-sessionTest` presents the JUnit test case that we have written for method `create-chat-session`. Initially, two users u1 and u2 are added to the set of users. Then, `run-create-chat-session(u1, u2)` creates a chat session for them. Class `Pair` implements a pair of elements.

- `assertTrue(users.has(u1) && users.has(u2));` validates that both users were actually added to `users`.

- `assertTrue(chatsessions.has(p12));` validates that a chat session between u1 and u2 is created.

- `assertFalse(chatsessions.has(p21));` validates that `create-chat-session(u1, u2)` does not actually create a chat session between u2 and u1.

```
public class create-chat-sessionTest {
  Integer u1, u2;
  machine2 m;

  @Before
  public void setUp() throws Exception {
   u1 = 1; u2 = 2;
   m = new machine2();
  }

  @After
  public void tearDown() throws Exception {
   u1 = u2 = 0;
   m = null;
  }

  @Test
  public void testCreateChat() {
   m.add-user(u1);
   m.add-user(u2);
   m.create-chat-session(u1,u2);

   BSet<Integer> users = m.get-user();
   BSet<Pair<Integer,Integer>> chatsessions = m.get-chat();

   Pair<Integer,Integer> p12 = new Pair<Integer,Integer>(1,2);
   Pair<Integer,Integer> p21 = new Pair<Integer,Integer>(2,1);

   assertTrue(users.has(u1) && users.has(u2));
   assertTrue(chatsessions.has(p12));
   assertFalse(chatsessions.has(p21));
  }

}
```

3.9   FIXING THE SOFTWARE REQUIREMENTS

In Section 2.4, we introduced a methodology for the early validation of software requirements. Therefore, USs are ported to a formal model in EVENT B, then the EVENTB2JAVA code generator is used to generate a prototype JAVA implementation of the EVENT B model, which is unit-tested to validate the USs and the formal model.

Our main goal here is to check US-03 (the chatting event functionality) for flaws. Therefore, we unit-test the JAVA code that EVENTB2JAVA generates for event chatting, we check if the code can be exercised (executed) properly and under which conditions, and, if necessary, we evolve the US and the EVENT B model. Right until now we have essentially encoded each and every US as an event. We bring below US-03, which has already been presented in Section 3.1. US-03 describes the chatting functionality through which "Me" chats with "You". According to US-03, chat content is made available to these two users.

US-03	chatting
chatting	As a user, I want to send You some content during a chat session with You so as to express my ideas
Acceptance criterion	Given: The chat session between Me and You is active When: I chat with You Then: The chat content is made available to Me as well as to You

We present below the whole structure for unit-testing the chatting functionality, for which we have written the testChatting() unit-test below.

```
public class chattingTest {
  Integer c1, c2, c3;
  Integer u1, u2, u3;
  machine2 m;

  @Before
  public void setUp() throws Exception {
   c1 = 1; c2 = 2; c3 = 3;
   u1 = 11; u2 = 12; u3 = 13;
   m = new machine2();s
  }

  @After
```

```
public void tearDown() throws Exception {
  c1 = c2 = c3 = u1 = u2 = u3 = 0;
  m = null;
}


// testChatting()


}
```

testChatting() creates two users Me and You, it also creates a chat session between them and makes Me send some content c1 to You. Index k1 is the (future final) position of c1 in Me's screen, and k2 is its position in You's screen.

```
@Test
public void testChatting() {
  m.add-user(Me);
  m.add-user(You);

  m.create-chat-session(Me, You);
  Integer k1 = 0, k2 = 0;

  m.chatting(c1, Me, You, k1, k2);
  assertTrue(m.get-screen().apply(Me).apply(You).domain().has(k1));
}
```

Not only testChatting() fails, but it also raises a JAVA null pointer exception. More concretely, in the last line of the testChatting() test, the expression m.get-screen().apply(Me) is empty, so m.get-screen().apply(Me).apply(You) is null, and hence m.get-screen(). apply(Me).apply(You).domain() raises a null pointer exception in JAVA. The problem is that the event guard You∈dom(screen(Me)) of event chatting does not hold (see guard @grdr21 of event chatting in Section 3.6) the first time that Me sends some content to You, and hence dom(screen(Me)(You)) is not well defined.

To remedy this problem, we edit US-03 and split it into the US-03a and US-03b USs for the cases when Me and You are chatting for the first time or not, respectively. US-03a and US-03b are translated to EVENT B, checked with Rodin, and all the generated POs are discharged. Event is similar to event chatting, except that it checks that the two users are chatting for the first time.

US-03a	chatting-first-time
chatting	As a user, I want to send You some content during a chat session with You so as to express my ideas
Acceptance criterion	Given: The chat session between Me and You is active Given: Some chat content for the chat session between Me and You exists as well as for the session between You and Me When: I chat with You Then: The chat content is made available to Me as well as to You

US-03b	chatting
chatting	As a user, I want to send You some content during a chat session with You so as to express my ideas
Acceptance criterion	Given: The chat session between Me and You is active When: I chat with You Then: The chat content is made available to Me as well as to You

chatting-first-time should be used the first time that Me sends some content to You, otherwise chatting should be used instead. However, none of the two events can be used if one of the following conditions occurs after Me has started chatting with You or vice versa.

- Me deletes her chat session with You so that You∉dom(screen(Me)) although Me∈dom(screen(You)).

- You deletes her chat session with Me so that Me∈dom(screen(You)) although You∉dom(screen(Me)).

Event chatting is expressed similar to chatting-first-time below, except that the second conjuncts of @grdr21 and @grdr22 are not negated.

```
event chatting-first-time extends chatting
any k1 k2
where
  @grdr21 Me∈dom(screen) ∧ You∉dom(screen(Me)) ∧ k1∈ℤ
  @grdr22 You∈dom(screen) ∧ Me∈dom(screen(You)) ∧ k2∈ℤ
then
  @actr21 screen ≔ screen ⊗
          {Me ↦ ( screen(Me) ⊗ { You ↦ {k1↦c}}),
              You ↦ ( screen(u2) ⊗ {Me ↦ {k2↦c}})}
```

```
@actr22 csize ≔ csize+1
@actr23 contents ≔ contents ⊗ {csize+1↦c}
end
```

We then generate code for the new versions of `chatting-first-time` and `chatting` and modify the `testChatting()` test to use `chatting-first-time` instead of `chatting`. The first `assertTrue` instruction checks that content `c1` is visible to Me, and the second that it is visible to You. Both tests succeed this time and hence US-03a and US-03b are correct.

```
@Test
public void testChatting() {
  m.add-user(Me);
  m.add-user(You);

  m.create-chat-session(Me, You);
  Integer k1 = 0, k2 = 0;

  m.chatting-first-time(c1, Me, You, k1, k2);
  assertTrue(m.get-screen().apply(Me).apply(You).domain().has(k1));
  assertTrue(m.get-screen().apply(You).apply(Me).domain().has(k2));
}
```

3.10   LESSONS LEARNED

3.10.1   EVENT `CREATE-CHAT-SESSION`

What if one adds the guard Me↦You ∈ `active` to the event `create-chat-session` in Section 2.9? After adding the aforesaid guard, Rodin discharges all the related POs automatically. The consequences of making or not Me↦You ∈ `active` are rather related to the `create-chat-session`'s interactions with other machine events such as `chatting` and `select-chat`. If `create-chat-session` does not make Me↦You ∈ `active` then `select-chat` should execute afterwards, before start chatting. The analysis of the event interaction of an EVENT B model can typically be performed using the PROB model checker (Leuchel and Butler, 2003). PROB checks for deadlocks: it checks if after executing any of the machine events, the system can make progress or not.

3.10.2 EVENT CHATTING

The abstract version of the event `chatting` is presented below (the version defined in `MachineZero`). Regarding this abstract version, the first decision we need to make is whether or not we want to add content item `c` to the chat between `You` and `Me` (in addition to the chat between `Me` and `You`). If we want to do so, we should extend the second line of `@act3` below with `You ↦` `(chatcontent(You) ∪ {c ↦ {Me}})` as we have done. Intuitively, adding this line means that the content `c` sent by Me to You is not only seen by Me but also by You. However, if we choose to extend `@act3` that way, Rodin provers would generate a PO henceforth `You` must be in the `dom(chatcontent)` so that sub-expression `chatcontent(You)` is well-typed. This requirement can easily be solved by adding an event guard `@grd6 You ∈ dom(chatcontent)` to the `chatting` event. The downside of this solution is that `@grd6` does not hold the first time when `You` has not yet sent any content to anyone (not just to `Me`).

```
@act3 chatcontent := chatcontent ⊕
        {Me↦(chatcontent(Me) ∪ {c↦{You}})} ⊕
        {You↦{c↦{Me}}}
```

The above downside suggests that one could add default `chat-content` associations the first time `You` is added to the system. The event `add-user` below adds user `u` to the set of existing users. Set expression `USER\user` returns the set of all the users that have not yet been added to the system. Action `@act2` adds default `chat-content` associations for user `u` with respect to the set of all the existing content. `∅` is the empty-set of users in this case. The soundness of `@act2` below is established by Rodin provers by discharging all the associated POs automatically; in particular, `@act2` adheres to `@inv4` in Section 3.4 which requires the definition to ensure a partial-function relationship between `users` and `content`.

```
event add-user
any u
where
 @grd1 u ∈ USER\user then
 @act1 user := user ∪ {u}
 @act2 chatcontent(u) := content × {∅}
end
```

What would happen with the association encoded by `@act2` above the next time that we create a new content item? Event `add-content` is shown below. For each and every existing user,

@act2 below associates the fresh content c to the empty-set of users, in other words, content item c appears as been sent by the whole set of users (variable user) to no one.

Summing up on the chatting event, if we wanted to add content item c to the chat You ↦ Me in addition to the chat Me ↦ You, then we would incur into a computationally expensive task: we would need to associate ø to every existing content item every time we add a user to the system, and we would need to associate every single user to {c ↦ ø} every time we needed to add a fresh content item c. This type of analysis on the complexity of associating chat content to You ↦ Me is not very intricate, in general; this analysis can be performed through careful code inspection or testing. But, writing the formal specification of our chatting system in EVENT B forces us to do code-inspection, and having Rodin theorem provers ensures that all cases are considered when performing automatic checking of EVENT B specifications with Rodin, without having to put effort into writing appropriate test scenarios.

Notice that expressing @act3 in chatting as below does not work since the last overriding expression forgets about chatcontent(You), which amounts to deleting it. This issue cannot be spotted by Rodin, in particular regarding invariant @inv4 in Section 3.4, as the new association for You below would be a partial function anyway. This can only be spotted by a domain expert who knows that deleting a person's chat whenever a content item is sent to her is an unwanted side effect.

```
@act3 chatcontent := chatcontent ⊕
        {Me ↦ (chatcontent(Me) ∪ {c ↦ {You}})} ⊕
        {You ↦ {c ↦ {Me}}}
```

The final solution is to use a set-comprehension expression to express the new value of chatcontent(You) as indicated in the last overriding expression below. The downside of this solution is that the expression cannot directly be encoded with sets, relations, and their operators (domain restriction, domain subtraction, inverse, etc.) by code generators such as EVENTB2JAVA.

```
@act3 chatcontent := chatcontent
    ⊕ {Me ↦ (chatcontent(Me) ∪ {c ↦ {You}})}
    ⊕ {cc,s · You ↦ {cc ↦ s} ∈ chatcontent ∨ (cc=c ∧ s={Me})
            | You ↦ {cc ↦ s}}
```

3.10.3 EVENTS DELETE-CONTENT AND REMOVE-CONTENT

delete-content and remove-content are two of the subtlest functionalities of our chatting system since that the performing of either can break invariants all around. delete-content below does not remove c from chat You ↦ Me. Under which circumstances should one add action @act2 content := content \ {c} to event delete-content? If we add @act2 to delete-content, Rodin will generate an unprovable PO. The PO relates to @inv4 in Section 3.3. One would need to demonstrate that for any user Someone other than Me the range of chatcontent(Someone) is a partial function from content\{c} to ℙ(user), which is not possible because it might be the case that Someone had sent (forwarded or broadcasted) c to another user previously.

```
event delete-content // US-04a
any Me You c
where
  @grd1 Me ∈ user ∧ You ∈ user
  @grd2 Me ↦ You ∈ active
  @grd3 Me ∈ dom(chatcontent)
  @grd4 c ∈ dom(chatcontent(Me))
  @grd5 You ∈ chatcontent(Me)(c)
then
  @act1 chatcontent(Me) := chatcontent(Me) ⊕
                            {c ↦ (chatcontent(Me)(c)\{You})}
end
```

A turn-around to this problem is to express content as below. However, to calculate the value of chatcontent that way one should traverse user twice and content once, which might be time-consuming depending on the type of structure used to store chatcontent or to represent sets in general.

```
@act2 content := {cc,a,b,s · a∈dom(chatcontent) ∧
                    cc ↦ s ∈ chatcontent(a) ∧
                    b∈s ∧ ¬(a=Me ∧ b=You ∧ cc=c) | cc}
```

Notice that if Me is chatting with You, and You with Another, and Me sends c to You, and You sends c to Another, calling remove-content with parameters Me, You, and c does not remove c from the chat between You and Another, but only from the chat between Me and You and between You and Me. For this reason, remove-content does not implement a second action @act2 content := content\{c}. To express the new value of content we can adopt the same approach as above and add the following line to the event remove-content.

```
@act2 content := {cc,a,s · a∈dom(chatcontent) ∧
                           cc ↦ s ∈ chatcontent(a) ∧
¬(a=Me ∧ cc=c) | cc}
```

CHAPTER 4

The Poporo Social Network

This chapter presents a complete EVENT B formal model of a social networking site called *Poporo*. A Poporo is a receptacle made of gold, used by pre-Columbian (before Christopher Columbus) Amerindians to socialize and communicate with ancient gods. Poporos were believed to have mystical and social communication powers. These mystical communication powers are arguably due to a mixture of quicklime and mashed coca leaves substances that were poured into the Poporos. Poporos belong to the Chibcha, Muisca, Quimbaya, and Calima pre-Columbian cultures.

Figure 4.1: Poporo Muisca.

Figure 4.1 presents a typical Muisca's Poporo. The presented social network is named Poporo in honor of these first Amerindian cultures that gathered and socialized together.

Section 2.3 presents a series of steps of software development methodology based on EVENT B. These steps offer a supplementary view of the early validation methodology presented in Section 2.4. We show those steps again below.

1. An initial blueprint (machine) of the system is written in EVENT B. The initial blueprint plays the role of the most abstract view of the system that we want to model.

2. Soundness proofs that demonstrate that each event in the initial blueprint adheres to the invariant properties are conducted, for instance, by using the Rodin platform.

3. A model refinement for the blueprint is written. You can refine your model horizontally, vertically or combine both.

4. Consistency proofs that demonstrate that the blueprint refinement is a correct refinement of the blueprint are conducted.

5. The two previous steps are iterated as many times as desired.

6. Code is implemented or tool generated for the final model refinement.

Whereas Chapter 3 is dedicated to steps 1 and 3, this chapter is rather dedicated to steps 2 and 4. This chapter shows how to discharge (prove) some typical POs for Poporo semi-automatically in Rodin. Rodin includes a plug-in called EVENTB2JAVA (Cataño et al., 2015) that generates JAVA code for EVENT B models automatically.

Table 4.1 shows a simplified model for social networking in EVENT B. Our complete formalization for social networking is called *Poporo*. It shares many features with existing well-known social networking web sites like Facebook and others, and additionally, all its components are formalized in predicate logic. The Poporo machine includes definitions for variables, invariants, and events. The Poporo machine *sees* the context ctx. The context includes definitions for constants, carrier sets, axioms, and theorems. Constants and variables represent the *static part* of an EVENT B model, whereas events constitute its *dynamic part*, and thus they might modify (machine) state variables. Thus, the Poporo machine can access all the constant definitions, axioms, and theorems defined in ctx as if they were declared locally within it. The machine declares four variables, namely, person, content, owner, and page. The variable person represents the set of all persons of the social network. The variable content represents the set of all the content items of the social network, for instance, pictures, text messages, and videos. A variable owner models who owns which content, and page stores the contents of each person's home page. Poporo declares four invariants. The first two invariants typeset variables person and content. Elements of these variables are taken from *carrier sets* PERSON and CONTENT, respectively.

The third invariant in Table 4.1 states that owner is a *total surjective* function from content to person. The total condition ensures that some person owns every content item, and the surjective condition ensures that every person owns some content item. The fourth invariant declares that page is a total surjective relation between content and person.

The initialization event gives initial values to the machine variables. Poporo initializes all the machine variables to the empty set, denoted ∅. The event upload of the social network uploads certain content c to the page of person p. Variables c and p are parameters of the event upload. Guards are logical predicates formed of constants, machine variables, sets, and event parameters.

Table 4.1: A simplified version of the Poporo abstract machine

```
machine Poporo sees ctx
variables person content owner page
invariants
 @inv1 person ⊆ PERSON
 @inv2 content ⊆ CONTENT
 @inv3 owner ∈ content ⇸ person
 @inv4 page ∈ content ⇸ person
events
 event initialization
 then
  @init1 person ≔ ∅ @init2 content ≔ ∅
  @init3 owner ≔ ∅ @init4 page ≔ ∅
 end


 event upload
 any c p
 where
  @grd1 c ∈ CONTENT \ content
  @grd2 p ∈ person
 then
  @act1 content ≔ content ∪ {c}
  @act2 owner ≔ owner ∪ {c↦p}
  @act3 page ≔ page ∪ {c↦p}
 end
end
```

upload's guards are labeled @grd1 and @grd2. The first guard typesets parameter c and the second one variable p. The first guard checks that c is a fresh content, and the second guard checks that p is an existing person. upload's actions are labeled @act1, @act2, and @act3. Each event action must assign to a different machine variable. The first action adds content item c to content. The second action adds the pair c↦p to owner. Thus, person p becomes the owner of the fresh content item c. The third action adds content item c to the page of person p. Event guards play the role of event *pre-conditions*. Events do not have an explicit notation for a variable *post-condition*, nevertheless, in EVENT B, the use of a variable on the right-hand side of an assignment denotes the value of the variable in the pre-state of the event, and its use on the left-hand side of an assignment denotes its value in the post-state of the execution of the event.

4.1   POPORO'S GENERAL STRUCTURE

Poporo's most abstract functionality is described in Table 4.2. Its third column tells which requirements are related to which EVENT B machine or refinement. This basic model considers an abstract machine and one single machine refinement. The single refinement machine sees permissions on content items on top of what the abstract machine already sees. Table 4.3 shows an extension of the basic model that is slightly discussed here. The extended model sees friendship relationships, the wall, and a chatting room.

Table 4.2: Poporo's initial model

Model	What does the model see?	Requirements
Poporo abstract machine	Users, content, owner	FUN-1, FUN-2, FUN-3
Refinement 1	Permissions	SEC-1, SEC-2, FUN-4, INV-1, INV-2

Table 4.3: Poporo's extended model

Model	What does the model see?
Refinement 2	Principal and secondary content
Refinement 3	Friendship relations (best friends, social friends, acquaintances)
Refinement 4	The wall
Refinement 5	The chatting room

Table 4.4 below summarizes the notation that we will be using in this chapter for writing software requirements. This notation differs from the notation used in Chapter 3 in which we use USs to write requirements. FUN is the set of functional requirements. SEC is the set of non-functional requirements related to security and privacy aspects of the Poporo social network. INV is the set of safety properties of Poporo. Each requirement is composed of two parts, the software requirement itself, and prose that conveys intended design decisions.

Table 4.4: Requirement's notation

FUN	Requirements about the functionality of the system
SEC	Requirements about privacy and security
INV	Invariant property

This separation between prose and requirement has two purposes. Prose helps new readers to get quickly acquainted with the system; it also helps frequent readers to read the requirements quickly. The requirement uses mathematical notation that is more suitable for expert readers. We write prose in a grey color.

In what follows, we present Poporo's abstract machine software requirements. Table 4.5 shows Poporo's functional requirements. FUN-1 states that the system should model users and data. The second requirement tells us that ownership should be modeled by the system state, and this requirement and the third one state that the system must dynamically be able to upload data. The fourth functional requirement describes a common feature in social network whereby content streaming can be made visible or invisible to someone's page. This requirement models the situation when one wants to switch on or off people's updates.

Table 4.5: Poporo's functional requirements	
FUN-1	The social network shall have users and data (content)
	Formalization of the system state
FUN-2	The user who uploads data shall be classified as the owner of the data
	Definition of ownership
FUN-3	The users of the social network shall upload data
	Publishing content on the web
FUN-4	Users might choose what data available to them is viewed by them
	Hiding and making content visible

Table 4.6 shows Poporo's privacy and security requirements. The first requirement tells us that Poporo's privacy and security mechanism is based on *access permissions*. The second requirement introduces the types of access permissions that Poporo encompasses, namely, *view* and *edit* permissions.

Table 4.6: Poporo's non-functional requirements	
SEC-1	The users shall have controlled access to the data on the network based on permissions
	Content privileges
SEC-2	The following permissions over a data may be given to a user:
	The permission to view the data
	The permission to edit the data
	View and edit permissions

Table 4.7 shows some of Poporo's safety properties. In EVENT B, safety properties are specified as machine invariants. The first safety requirement imposes a hierarchy between the two considered access permissions. The second safety property tells that a person that uploads a con-

tent item, and hence owns it, possesses all the permissions on that content.

Table 4.7: Poporo's safety properties	
INV-1	Users that can edit data must also be able to view it
	Edit permissions subsume view permissions
INV-2	The owner of some data has all the permissions on it
	Definition of ownership

4.2   POPORO'S FORMALIZATION IN Event B

Context `ctx` below defines the carrier sets for the persons and the content items in the social network. `PERSON` and `CONTENT` are uninterpreted sets, and hence they're not constrained to contain elements of any particular type.

```
context ctx
 sets PERSON CONTENT
end
```

FUN-1 is about a static aspect of social networks that has to do with users and content. We use `person` to model users, and `content` to represent social network content. `person` and `content` are subsets of the two carrier sets defined in `ctx`. The label used in each invariant below help PO generators to map the invariant to a particular PO.

```
variables person content
invariants
 @inv1 person ⊆ PERSON
 @inv2 content ⊆ CONTENT
```

FUN-2 is related to the ownership of social network content. One can think of several possible definitions for the ownership of content; the one we selected here involves network content and people. One can define `owner` as a relation between `content` and `person`, but this would allow, for instance, several persons to own the same network content, which is not what we want to model here. One should thus model `owner` as a function. If we model owner as a *partial function* from `content` to `person`, then we would permit some content to exist that is not owned by anyone in the social network, which does not comply with the typical behavior social networks have. Therefore, we choose to model owner as a *total function*. We also choose owner to be *surjective*. This ensures that each person has at least one owned content item. We show the definition of `owner` below.

```
@inv3 owner ∈ content ⇸ person
```

We use a fresh label for our third machine invariant. As said before, labels enable automated tools to generate distinctive POs. Section 4.4 presents an example of how an invariant PO is proven using Rodin.

The static part of FUN-3 involves page content and people. We use a relation `page` from `content` to `person` to model page content (see `@inv4` below). `page` is a *total surjective* relation. It associates every element in its range with at least one element in its domain (surjective). It further associates each content item in its domain with at least one person in its range (total). Therefore, each content item is in someone's page, and every person has at least one content item on their page.

We could have defined `page` as a relation from `person` to `content` instead. However, we want our definitions to be as uniform as possible. For instance, `page`'s definition is uniform with respect to `owner`'s definition; they share the same domain and range types. Uniformedness is useful when one has to compose different relations, or state properties of the inverse of some relation with respect to another, or state a set inclusion property between two relations.

```
@inv4 page ∈ content ⤀ person
```

FUN-3 also involves a dynamic part. It tells about the user's ability to upload content to the network. Dynamic behavior is implemented with the aid of events. The event `upload` below implements the dynamic part of FUN- 3. The event takes two parameters, `c` and `p`. The two event guards are the typing conditions for the two parameters, respectively. `@grd1` checks that `p` is an existing person, and `@grd2` checks that `c` is a fresh content. Event `upload` includes three actions. In EVENT B, event actions must assign into a set of different variables, and the execution of the event actions together are assumed to take no time; they are *atomic*. The three actions modify `content`, `owner`, and `page`, respectively.

```
event upload
any c p where
 @grd1 p ∈ person
 @grd2 c ∈ CONTENT\content
then
 @act1 content ≔ content ∪ {c}
 @act2 owner ≔ owner ∪ {c↦p}
 @act3 page ≔ page ∪ {c↦ p}
end
```

4.3   INVARIANT POs

A PO describes what should be proven for a model to be correct. Proof generators and proof assistants automatically generate POs. A PO is a *sequent* formed of *hypotheses* and a *goal*. Of particular interest for us are *invariant proof obligations*. The execution of machine events must comport with machine invariants. Hence, for each event definition, invariant proof obligations are generated that testify whether or not events adhere to machine invariants, that is, whether the execution of the event invalidates the machine invariants or not. Invariant POs must be discharged to ensure the correctness of machine events.

$I(s, c, v)$	Machine invariant
$G(s, c, v, x)$	Event guard
$\vdash$	INV
$I(s, c, E(s, c, v, v', x))$	Modified invariant

Figure 4.2: Invariant PO rule.

Rule INV shows the invariant PO rule for events. $I(s, c, v)$ is the machine invariant, which depends on all the sets s and constants c defined in the machine context, and the machine variables v. Event guard G additionally depends on the set of event parameters x. One must prove that the *modified invariant* $I(s, c, E(s, c, v, v', x))$ holds. The modified invariant accounts for the effect produced by the execution of the event actions, hence, one must prove that the machine invariant I holds after replacing the set of machine variables v by expression E. Notation v' is used for the value of v after the execution of the event, that is, the value of v in the post-state.

Which POs should one discharge to ensure that event upload complies with invariant @inv3? Invariants are assumed to hold before the execution of the event and must be proven to be correct after its execution. Invariant @inv3 states that owner is a total surjective function. One must prove that if owner was a total surjective function before event upload executes, then it continues to be a total surjective function after it executes. One must prove that owner, after adding the pair c↦p to it, is a total surjective function that goes from the modified content to person.

To discharge this PO, one must reason for parts. The modified owner is a function since it was a function before the event executes, and the fresh content c is not in its domain. The modified owner is a total function since it is a function, it was initially total, and it sets an image for a new element added to its domain. The modified owner is surjective because it was surjective before the event executes, and the pre-image of p is c. Section 4.4 discusses how invariant POs can be discharged (conducted) with a proof assistant.

```
owner ∈ content ⤀ person
p ∈ person
c ∈ CONTENT\content
⊢
owner ∪ {c↦p} ∈ (content ∪ {c}) ⤀ person
```

Figure 4.3: Invariant PO for owner.

4.4 DISCHARGING POs IN RODIN

Figure 4.4 shows the event delete as defined in Poporo's abstract machine in Rodin. An existing content item c, a set of content items cts, and their owner ow parameterise the event delete. The event deletes c (*principal content*) and cts (associated *secondary content*). The bottom right part of the figure shows the Symbols toolbox for writing standard EVENT B mathematical symbols. Discharged (proven) POs are depicted in green (left-hand toolbox), undischarged in red, and unattempted POs in brown color. Rodin uses different *Perspectives*. The upper-right part of the figure shows that the current perspective is the *Event B perspective*.

One can either manually switch to a particular perspective (*Window, Open Perspective*) or undertake an action that makes the Rodin IDE switch to a default perspective for the performed action. If we double-click a PO (shown on the left panel), then it is open in the *Proving Perspective* of Rodin. Rodin comes with a list of predefined provers. A particular prover may or may not succeed to discharge a particular PO, and hence one is forced to use a different prover or to attempt to discharge the PO manually.

The bottom part of Figure 4.4 shows the various provers shipped with Rodin[6]. The PP prover is a third-party prover from the AtelierB tool. nPP is a native prover to Rodin. A *lasso* operation adds more hypotheses to the current proof that share common variables between the goal and the selected hypotheses. Provers can be used with different forces. p0 (PP with force 0) feeds the PP prover with the selected hypotheses and with the current goal. p1 (PP with force 1) performs a lasso operation before passing the hypotheses and goal to PP. nPP can be used with different forces too. The Rodin manual recommends the use of nPP before attempting any proof with PP. The ML prover is useful to discharge proofs that involve arithmetic.

---

[6] To install the most basic Rodin provers, on needs to add the Update Site http://methode-b.com/update_site/atelierb_provers.

Figure 4.4: Rodin IDE.

Next, We show how to discharge a PO in Rodin. The example is an invariant PO for event delete. The PO is generated to attest compliance of event delete to machine invariant @inv4 in Section 4.2. The proof is conducted with Rodin version 3.2.

```
1.    page ∈ content ↔ person
2.    c ∈ content
3.    {c} ⊂ dom(owner ▷ {ow})
4.    cts ⊆ content
5.    owner[{c} ∪ cts] = {ow}
      ⊢
Goal ({c} ∪ cts] = content\({c} ∪ cts) ↔ person
```

Figure 4.5: @inv4 invariant PO for event delete.

nPP and PP both fail to discharge the PO automatically. We adopt the following approach to discharge this PO. We strengthen the hypotheses, and then we undertake the proof by parts. Strengthening the hypotheses boils down to proving an intermediate result. Undertaking a proof by parts boils down to applying the divide-and-conquer principle. If we want to prove that some relation is *total surjective*, we first prove that it is *total*, then we prove that it is *surjective*, and finally we build on these two independent proofs.

The command ah of Rodin adds a new hypothesis to the list of hypotheses of a proof. The hypothesis, an auxiliary result, must be proven correct first. We add the auxiliary result below, a weaker version of our main goal above. nPP discharges the auxiliary result automatically. The result is hence added to the main goal as a sixth hypothesis.

$$({\{c\}} \cup cts) \triangleleft \, page \in content \backslash ({\{c\}} \cup cts) \leftrightarrow person$$

After proving the previous auxiliary result, we split the main proof into two parts; we first prove the *total* part of the goal and then the *surjective* part. We add the hypothesis below that accounts for the total part of the main goal.

$$({\{c\}} \cup cts) \triangleleft \, page \in content \backslash ({\{c\}} \cup cts) \leftrightarrow\!\!\!\!\leftrightarrow person$$

nPP and PP both fail to discharge this hypothesis automatically, and hence we decided to help the prover do its job by adding a yet simpler hypothesis on top of the previous one, which nPP discharges straightforwardly. The result below is added as the seventh hypothesis of the proof of the hypothesis above. Calling nPP with selected hypotheses sixth and seventh henceforth discharges the result above.

$$dom(({\{c\}} \cup cts) \triangleleft \, page) = content \backslash ({\{c\}} \cup cts)$$

We add the hypothesis below that accounts for the surjective part of the main goal, which again nPP and PP both fail to discharge.

$$({\{c\}} \cup cts) \triangleleft \, page \in content \backslash ({\{c\}} \cup cts) \leftrightarrow\!\!\!\!\rightarrow person$$

We hence add the following hypothesis, which is proven with PP. The hypothesis above is discharged with nPP by selecting the sixth hypothesis and the hypothesis below.

$$ran(({\{c\}} \cup cts) \triangleright \, page) = person$$

Surprisingly, after having proven the total and surjective parts, Rodin provers still fail to discharge the main proof. The problem is caused by the expression {c} ∪ cts, which Rodin attempts to simplify. We eliminate the noise caused by that expression with the aid of the command ae (added expression). We tell Rodin to see {c} ∪ cts as a whole indivisible expression. After that, nPP discharges the main goal with the selected hypotheses.

4.5   POS FOR QUANTIFIED EXPRESSIONS

Discharging POs that involve universally or existentially quantified properties is often complex since provers cannot instantiate quantifiers automatically, in general. We show a small example of how a universally quantified invariant PO is proven in Rodin. We model an invariant property that says that people can only own content items that exist in their web pages, and we present two equivalent versions of that invariant: invariant @inv51 uses set operations; and invariant @inv52 uses a universally quantified expression.

```
variables person content
invariants
  @inv51 owner ⊆ page
  @inv52 ∀c1, p1·c1∈content ∧ p1∈person ⇒
            (c1↦p1 ∈ owner ⇒ c1↦p1 ∈ page)
```

We are interested in discharging the PO generated for invariant @inv52 with respect to event hide below, which hides a content item c from the page of a user p.

```
event hide
any c p
where
  @grd1 p ∈ person
  @grd2 c ∈ content
  @grd3 c↦p ∈ page
  @grd4 owner(c) ≠ p
then
  @act1 page ≔ page\{c↦p}
end
```

Following up the invariant PO rule in Figure 4.2 in Section 4.3, a *sequent* is formed with hypotheses that account for @inv52 and the guards of the event hide, and the goal of the sequence is a modified version of @inv52 that accounts for the state changes introduced by the execution of the event actions. Figure 4.6 shows the sequent. Hypothesis 1 is invariant @inv5 2, hypotheses 2 to 5 are the event guards. Rodin automatically instantiates the goal (a universally quantified invariant) and applies modus ponens, hence the hypothesis number 6. The latter hypothesis typesets c1 and p1.

```
1.     ∀ c1, p1 · c1 ∈ content ∧ p1 ∈ person ⇒
                    c1↦p1 ∈ owner ⇒ c1↦p1 ∈ page
2.     p ∈ person
3.     c ∈ content
4.     c↦p ∈ page
5.     owner(c) ≠ p
6.     c1↦p1 ∈ owner
       ⊢
Goal  c1↦p1 ∈ page\{c↦p}
```

Figure 4.6: PO for event hide as per @inv52.

To discharge this PO one should instantiate the first hypothesis with appropriate values. One needs two values of type content and person such that the mapping from the latter to the former is an element of owner. After the instantiation of this hypothesis with c1 and p1, nPP discharges the PO automatically. However, had we defined @inv52 as @inv51, the prover would have discharged the underlying PO without human intervention. One should avoid writing invariant properties that use quantifiers in general, and foresees ways those can be expressed using set and relation operations instead.

4.6   STRENGTHENING THE SPECIFICATION

This section partially shows the development of the first refinement machine in Table 4.2. The refinement involves requirements SEC-1, SEC-2, FUN-4, INV-1, and INV-2. SEC-1 and SEC-2 refer to the static part of the refinement machine; they extend Poporo's abstract machine to model access permissions on network content. We use two relations viewp and editp as below to represent *view* and *edit* permissions. Note that labels in the refinement machine are different to the labels in the refined machine. This allows tools like Rodin to generate POs for invariants.

We could have defined viewp and editp from person to content instead, but then we want definitions for viewp and editp to be uniform to the definition for page in the abstract machine, and therefore to be able to use operators like ⊂ or ⊆ to compare page content and permissions on content directly.

```
variables viewp editp

invariants
  @inv11 viewp ∈ content ↔ person
  @invr12 editp ∈ content ↔ person
```

INV-1 introduces a static hierarchy between the two access permissions, and INV-2 describes a safety property of the system whereby the owner of a content item may access it in any possible way.

```
variables viewp editp

invariants
  @invr13 editp ⊆ viewp
  @invr14 owner ⊆ viewp
  @invr15 owner ⊆ editp
```

Refinement machines must keep a palpable behavioral relationship with its abstraction (abstract machine). This relationship is modeled through a *gluing invariant* property that relates variables of the refinement machine with variables of the refined machine. Roughly speaking, a refinement model should be such that it can replace the refined (abstract) model without the user of it noticing any change. Next, we describe some properties about gluing invariants.

- Gluing invariants are defined in the refinement machine.

- Gluing invariants relate the state of the refinement machine to the state of the refined machine.

- Gluing invariants must be preserved by the events of the refinement machine, therefore, the gluing invariant generates POs that the refinement machine must verify.

- As a refinement event refines or extends an abstract (refined) event,[7] concrete events must comply with invariants of the refined machine reachable through the gluing invariants.

---

[7] Details on refines and extends are given in Section 4.8.

We introduce a gluing invariant that relates view permissions in the refinement machine with page content in the refined machine: every view permission on a person's content is defined over the existing content in her page. Note that one does not need to add a similar invariant that relates edit permissions and page content; this invariant would be inferred from invariants `@invr13` and `@invr16`.

```
invariants
  @invr16 viewp ⊆ page
```

FUN-4 refers to the dynamic part of the refinement machine, for which we write a `grant-view-permission` event. `@grd1` and `@grd2` typeset content `c` and the person `p` to whom the permission will be granted. `@grd3` expresses a defensive style of writing specifications: We want to grant permissions to content items that do not have those permissions already. The job of `@grd4` is to preserve `@invr16`. As we are adding an element to `viewp`, either we add the same element to `page`, or we require the element to exist in `page`, as we have.

```
event grant-view-permission
any pc where
  @grdr11 p ∈ person
  @grdr12 c ∈ content
  @grdr13 c↦p ∉ viewp
  @grdr14 c↦p ∈ page
then
  @actr11 viewp ≔ viewp ∪ {c↦p}
end
```

Poporo implements some functionality that deletes network content. Event `delete` deletes a content item `c` and all the content items in `cts` from all the pages in the network. The intended meaning of `cts` will be clearer when we refine the observation about the system (see Section 4.7). `@grdr11` typesets `c`. `@grdr12` prevents the event from leaving `owner(c)` with an empty page. `@grdr13` typesets `cts`.

Section 4.7 shows the observations of principal and secondary content. Section 4.8 shows what POs are generated for further observations (refinements) to be correct.

```
event delete any c cts
where
```

```
    @grdr11  c∈content
    @grdr12  {c} ⊂ dom(owner▷{owner(c)})
    @grdr13  cts ⊆ content
    @grdr14  c∉ cts
  then
    @actr11  content ≔ content \ ({c} ∪ cts)
    @actr12  owner ≔ ({c} ∪ cts) ◁ owner
    @actr13  page ≔ ({c} ∪ cts) ◁ page
  end
```

4.7   FURTHER STRENGTHENING

We add the observation of page fields. Each content item belongs to some field. Content fields model the typical situation in social networks in which people comment on certain content-item. The comment is a field or secondary content item and the commented content item is a principal content item. A field belongs to a unique principal content; the same way a photo belongs to a unique photo album. Deleting a principal content item entails deleting all its fields, the same way that deleting a photo album would entail deleting all its photos, or deleting a content item would force one to remove all the content item comments too. Its owner can only remove a principal content item. Expression field~[{c}] returns all the *secondary* content items whose principal content item is c. Definitions for principal and secondary content are given next.

```
    variables  principal field


    invariants
    @invr17  principal ⊆ contents
    @invr18  field ∈ contents\principal → principal
```

Events can be refined by using either a refines or an extends keyword. These two mechanisms are alike. An extending event implicitly includes parameters, guards, and actions of the extended event; a refinement event does not. A refinement event can additionally declare a *witness* (using the clause with) for every disappearing parameter of the abstract event. A witness for a parameter x in the refined (abstract) event is a property P(x) declared in the refinement (concrete) event, typically written: "with @x P(x)."

```
event delete-principal refines delete
any c
where
 @grdr21 c ∈ principal
 @grdr22 owner[principal\{c}] = persons
with
 @cts cts = field~[{c}]
then
 @actr21 content ≔ content \ ({c} ∪ field~[{c}])
 @actr22 owner ≔ ({c} ∪ field~[{c}]) ◁ owner
 @actr23 page ≔ ({c} ∪ field~[{c}]) ◁ page
 @actr24 viewp ≔({c} ∪ field~[{c}]) ◁ viewp
 @actr25 editp ≔ ({c} ∪ field~[{c}]) ◁ editp
 @actr26 principal ≔ principal \ {c}
 @actr27 field ≔ field ◁ {c}
end
```

Event `delete-principal` refines event `delete`. Parameter `cts` is a disappearing abstract parameter which we model in the concrete event as `field~[{c}]` with a `with` clause. Therefore, parameter `c` is a principal content and `cts` are all its fields. The three actions of the abstract event are repeated in the concrete event, yet not show explicitly. The fourth and fifth actions in the concrete event are written not to break invariants `@invr13` and `@invr16`. Actions `@actr26` and `@actr27` remove `c` and all its fields.

The same event can be modeled in a more succinct way by using an `extends` keyword as shown below. Nevertheless, a `refines` keyword offers more possibilities for (data) refinement. Had we defined a concrete machine gluing invariant that relates either `content`, `owner`, or `page` with the state of the abstract machine, we would not have needed to repeat the first three actions of the refined event, but instead we could have expressed them by using the gluing invariant.

The definition of refinement events generates POs for the adequacy and correctness of the events. Section 4.8 explains the POs that are generated for event `delete-principal` and shows how they can be proven.

```
event delete-principal2 extends delete
 @actr24 viewp ≔ ({c} ∪ field~[{c}]) ◁ viewp
 @actr25 editp ≔ ({c} ∪ field~[{c}]) ◁ editp
 @actr26 principal ≔ principal \ {c}
```

```
    @actr27 field ≔ field ◁ {c}
  end
```

4.8   REFINEMENT PROOF OBLIGATIONS

Refinement POs are generated to attest about the correctness of refinement events. The refinement invariant PO rule in Figure 4.7 is similar to the INV rule depicted in Figure 4.2, except that rule in Figure 4.7 accounts for both abstract and concrete machine invariants. The concrete machine invariant J depends on both the abstract and concrete machine variables v and w, respectively. H is the concrete event guard, and y is the set of concrete event parameters.

$I(s, c, v)$	Abstract machine invariant
$J(s, c, v, w)$	Concrete machine invariant
$H(s, c, w, y)$	Concrete event guard
$With(s, c, v, w, w', y)$	Witness predicate
$\vdash$	INV
$J(s, c, v', w')$	Modified concrete machine invariant

Figure 4.7: Invariant PO for refinement events.

Invariant POs are generated with respect to a particular event action that might invalidate the invariant. For instance, Rodin generates the PO `delete-principal/invr18/INV` below for event `delete-principal` with respect to invariant `@invr18` and even action `@actr27`. The first hypothesis below is the concrete machine invariant. The second and third hypotheses are the concrete event guards. The goal is the modified concrete machine invariant. The witness predicate about `cts` is expanded (unfolded) in the goal. Abstract machine invariants are not included explicitly as hypotheses but unfolded in the goal tool.

The *lasso* functionality of Rodin allows one to see all the hypotheses generated as explained in Section 4.4. We do not show the other concrete machine invariants either, as they are not needed to discharge the PO.

```
1.    field ∈ content\principal  ⤖  principal
2.    c ∈ principal
3.    owner[principal\{c}] = person
      ⊢
Goal  (field ⩥ {c}) ∈ (content\({c} ∪ field˜[{c}])) ⤖ principal\{c}
```

Figure 4.8: Invariant PO for delete-principal.

PO rule GRD below attests to the obligation for the abstract event to be enabled whenever the concrete event is. The concrete event guard must be stronger than the abstract event guard.

$I(s, c, v)$	Abstract machine invariant
$J(s, c, v, w)$	Concrete machine invariant
$H(s, c, w, y)$	Concrete event guard
$With(s, c, v, w, w', y)$	Witness predicate
⊢	GRD
$G(s, c, v, x)$	Abstract event guard

Figure 4.9: Guard PO for event refinement.

For the concrete event[8] delete-principal, and for the abstract event delete and the abstract event guard @grdr12, Rodin generates the following GRD PO as delete-principal/ grdr12/GRD. The two first hypotheses below are the concrete event guards, and the goal is the abstract event guard. We have intentionally omitted other hypotheses here. The second hypothesis says that owner(c) owns principal content other than c. This hypothesis directly entails the goal. Notice that @grdr11 is entailed by @grdr21 and @invr17, @grdr13 is entailed by @invr18, and @grdr14 is entailed by @invr18 and @grdr11.

```
1.    c ∈ principal
2.    owner[principal\{c}] = person
      ⊢
Goal  {c} ⊂ dom(owner ⩥ {owner(c)})
```

Figure 4.10: Guard PO for event delete-principal.

---

[8] "Concrete event" means the event in the concrete (refinement) machine, and the "abstract event" means the event in the abstract (refined) machine.

PO rule `SIM` ensures that the actions of the abstract event are correctly simulated by the actions of the concrete event. Concrete actions must not contradict the abstract ones, they must be stronger.

$I(s, c, v)$	Abstract machine invariant
$J(s, c, v, w)$	Concrete machine invariant
$H(s, c, w, y)$	Concrete event guard
$With(s, c, v, w, w', y)$	Witness predicate
$F(s, c, v, v', w, w', x)$	Concrete event actions
$\vdash$	SIM
$E(s, c, v, v', x)$	Abstract event actions

Figure 4.11: Simulation PO.

Actions of event delete are entailed by actions in the event `delete-principal` as we selected to repeat the actions of the former within the latter. Rodin generates other types of POs that we do not mention here, for instance, feasibility POs that ensure that proper values exist that make the concrete event guard true.

CHAPTER 5

Conclusion

The cost of fixing software design flaws after the completing of a software product is so high that it is vital to come up with ways to detect software design flaws in the early stages of software development before system coding starts. This is the motivation of this book as oftentimes software requirements are ambiguous, hence they contradict each other. This ambiguity is exacerbated by the fact that software requirements are typically written in natural language, which is not tied to any formal semantics. This book uses a palliative idea to deal with the ambiguity of software requirements: it restricts the syntax of software requirements to User Stories, textual templates with placeholders. However, as informal requirements (User Stories) do not enjoy any particular semantics, no essential properties about them (or about the system they attempt to describe) can be proven easily. Therefore, requirements are ported into EVENT B where we undertake mathematical proofs in Rodin to check their soundness.

This book is oriented to software developers who seek to increase their background in formal methods to develop software based on mathematical utilities. The book is divided into three main parts. The first is about Event B—a formal methods language that is based on discrete mathematics and predicate calculus. The second part of the book presents a case study on the software development of a chat system in EVENT B. The chat is composed of three main models that stem from an abstract functionality to a code related functionality. We unit-tested the code generated for the chat system and discovered several inconsistencies in the original model that we corrected with the aid of the Rodin platform and its underlying provers. The third part of the book presents the Poporo social network. This part does not focus on software development but on the formal proof of invariant properties.

We use a systematic approach for the early validation of software requirements which is based on the extensive use of formal methods techniques. This validation approach works on top of the refinement calculus mechanism of EVENT B and employs correctness-by-construction and deductive proof techniques with the RODIN platform. In short, software requirements are modeled as USs, then ported to EVENT B specifications, whose specification is refined as desired within the Rodin platform where generated POs are discharged. From the EVENT B formal specification, code is generated, simulated, and validated for flaws. Two types of validations are carried out. First, a sanity check of the formal specification is carried out to validate if the generated code can be used in standard scenarios. This process can result in the definition of new events. It can also result in the redefinition of event guards or the introduction of additional invariants. Second, code is generated from the formal specification. One can then write unit tests to simulate the generated code, and validate it against the informal requirements (the USs).

From a research and experimental viewpoint significant challenges and questions rest to be addressed to make the type of work presented in this book more cost-effective.

- How can we (semi-) automatically port USs to EVENT B syntax?

- How can we improve Rodin provers so that the rate of POs discharged automatically is increased?

- How can we generate unit tests automatically (from EVENT B specifications or code)?

The success of formal methods largely depends on the implementation of tools that assist users in the use of mathematical techniques that increase their trust in the software users want to develop. The EVENTB2JAVA code generator that's used in this book helped us increase our trust in the software requirements of the chat system, checking inconsistencies in the software requirements specifications that otherwise, through code inspection, would be impossible to check. The EVENT-B2JAVA tool does not generate optimal code but code that is correct. It is future work to improve the implementation of the tool to generate code that can serve as final system implementation. In particular, logical expressions in EVENT B that appear repeated within the definition of an event are translated and evaluated repeatedly in JAVA.

One major frustration in our work is the inadequate tool support for verifying JAVA programs with respect to the JML specifications that are generated by EVENTB2JAVA. Existing verification tools such as KeY and Krakatoa cannot handle the full syntax of JAVA and JML, particularly with regard to generics. We would like to undertake a case study on replacing parts of the code generated by EVENTB2JAVA with bespoke implementation and then verifying those implementations against the generated JML specifications. However, performing such verification without adequate tool support is time consuming and prone to error.

EVENTB2JAVA translates EVENT B axioms as JML invariants. These axioms determine the possible values that constants can take. However, EVENTB2JAVA cannot (yet) automatically generate values for constants that satisfy the axioms. We plan to investigate translating the EVENT B definitions of constants and axioms to the input language of the Z3 SMT solver, and then use Z3 to find values for the constants.

Bibliography

Abrial, J.-R. (2009). Faultless systems: Yes we can! *IEEE Computer*, 9(42), 30–36. DOI: 10.1109/MC.2009.283. 15

Abrial, J.-R. (2010). *Modeling in Event B: System and Software Design*. New York: Cambridge University Press. DOI: 10.1017/CBO9781139195881. 14

Abrial, J.-R. and Hallerstede, S. (2007). Refinement, decomposition, and instatiation of discrete models: Application to Event B. *Fundamenta Informaticae*, 77, 1–28. 14, 16

Abrial, J.-R., Butler, M., Hallerstede, S., Son Hoang, T., Mehta, F., and Voisin, L. (2010). Rodin: An open toolset for modeling and reasoning in Event B. *Software Tools for Technology Transfer*, 12, 447–466. DOI: 10.1007/s10009-010-0145-y. 1, 5, 16, 19

Almeida, J. B., Frade, M. J., Pinto, S. P., and Melo de Sousa, S. (2011). *Rigorous Software Development: An Introduction to Program Verification*. Porto: Springer Publishing Company, Incorporated. DOI: 10.1007/978-0-85729-018-2. 16

Appel, F. (2015). *Testing with JUnit*. New York: Packt Publishing. 49

Back, R.-J. and Sere, K. (1991). Stepwise refinement of action systems. *Structured Programming*, 12, 17–30. 3

Breunesse, C.-B., Catano, N., Huisman, M., and Jacobs, B. (2005). Formal methods for smart cards: An experience report. *Science of Computer Programming*, 55, 53–80. DOI: 10.1016/j.scico.2004.05.011.

Cataño, N. and Nishi, S. (n.d.). Soundness proof of EventB2Java. *Seventh Latin- American Symposium on Dependable Computing (LADC)* (pp. 25–34). Cali, Colombia: IEEE Digital Library. 18

Cataño, N. and Rivera, V. (2012). The EventB2Dafny Rodin Plugin. *Workshop on Developing Tools as Plug-ins (TOPI)* (pp. 49–54). Zurich: IEEE Xplore. DOI: 10.1109/TOPI.2012.6229810. 18

Cataño, N. and Rivera, V. (2016). Event2Java: A code generator for Event-B. *NASA Formal Methods (NFM)*. 9690, pp. 166–171. Minneapolis: LNCS. DOI: 10.1007/978-3-319-40648-0_13. 1, 15, 16, 43

Cataño, N., Rivera, V., Wahls, T., and Rueda, C. (2015). Code generation for Event-B. *International Journal on Software Tools for Technology Transfer*, 1–22. DOI: 10.1007/s10009-015-0381-2. 62

Cataño, N., Rivera, V., Wahls, T., and Rueda, C. (2017). Code generation for Event B. *International Journal on Software Tools for Technology Transfer*, 31–52. DOI: 10.1007/s10009-015-0381-2. 15, 16

Hall, A. (2002). Correctness by construction: Integrating formality into a commercial development process. *Formal Methods Europe* (pp. 224–233). Berlin: Springer Berlin Heidelberg. DOI: 10.1007/3-540-45614-7_13. 15, 18

Hall, A. and Chapman, R. (2002). Correctness by construction: Developing a commercial secure system. *IEEE Software*, 19, 18–25. DOI: 10.1109/52.976937. 15, 18

Leuchel, M. and Butler, M. (2003). ProB: A model checker for B. *FME: Formal Methods* (pp. 855–874). Pisa: Springer-Verlag. DOI: 10.1007/978-3-540-45236-2_46. 56

Index

Author's Biography

Néstor Cataño Collazos is a software engineer, computer scientist, and enthusiastic formal methods tool developer. His main research area is the use of formal methods for software engineering. Néstor specializes in program specification and verification using JML and design-by-contract, and in a formal method called EVENT B. His main goal is to build Formal Methods tools that increase people's trust in the correct behavior of Software Systems. His main tool contributions are the design and implementation of the EVENTB2JAVA and EVENTB2JML code generators, the design and implementation of the EVENTB2DAFNY Proof Obligation generator, and the design and implementation of the Chase syntactic checker of JML's assignable clause. Néstor's research work with EVENTB2DAFNY was funded by Microsoft Research through the SEIF program in 2011. He has worked in academia since 2004 and regularly teaches software engineering, formal methods, and programming courses to graduate and undergraduate students.

Dr. Collazos earned a Master Degree and a Ph.D. in Computer Science from The University Paris 7. He is currently taking part in the Master in Information and Cybersecurity at UC Berkeley.